THE POLITICS OF AGRICULTURAL MECHANIZATION IN CHINA

BY THE SAME AUTHOR

Making Green Revolution: The Politics of Agricultural Development in China
People's Communes and Rural Development in China
Rural Local Governance and Agricultural Development in Taiwan
We Were the Campaign: New Hampshire to Chicago for McCarthy

THE POLITICS OF AGRICULTURAL MECHANIZATION IN CHINA

Benedict Stavis

CORNELL UNIVERSITY PRESS

ITHACA AND LONDON

This book has been published with the aid of a grant from the Hull Memorial Publication Fund of Cornell University.

First published 1978 by Cornell University Press.
Published in the United Kingdom by Cornell University Press Ltd.,
2-4 Brook Street, London W1Y 1AA.

International Standard Book Number 0-8014-1087-8
Library of Congress Catalog Card Number 77-90916
Printed in the United States of America by Vail-Ballou Press, Inc.
Librarians: Library of Congress cataloging information appears on the last page of the book.

Agricultural mechanization is not merely a way of increasing production, but is part of a broad rural policy. It helps consolidate socialism in the rural areas, helps develop economic forces, helps consolidate the worker-peasant alliance, and helps eliminate the difference between the city and countryside.

—Li Kuan-heng, Tang Shan Special District Cadre,
responsible for agricultural mechanization, April 21, 1972

FOR MY PARENTS

Preface

The honeymoon with technology is over. The world has long thought that technology could solve most problems. Now, we know better. The multiple crises of the 1970's—food, energy, pollution, nuclear proliferation—demonstrate with frightening clarity the limitations of technology. But we cannot discard technology, either. Our civilization has been married to it for centuries; divorce would require the complete restructuring of the personality of our global culture.

In a broad sense we must start communicating again with technology. We must learn more clearly its capabilities and its limitations. We can no longer shy away from the political dimensions of technological choice. We have to realize that technology is embedded in particular social institutions, and the potentials and limitations of technology are shaped as much by politics as by technology per se.

This book does not attempt any global solution. It is simply a detailed analysis of how one country has faced the political and social dimensions of one particular technology. In a broad sense, the conclusions of this case study are encouraging. Through conscious political choice a program of agricultural mechanization is being developed in China in a manner that contributes to greater food production, more equitable food distribution, and significant job enrichment. The negative consequences of agricultural mechanization which are common elsewhere—unemployment, urban flight—are being avoided. Extrapolation into the future, however, involves uncertainties. China's policies are shaped by political choice, and political forces and alignments are always in flux as the very process of economic development creates new groups and interests and renders older ones irrelevant (but not impotent).

One of the recurrent themes of this book is the difficulties China has had precisely because it began by borrowing policies from the Soviet Union. What China eventually did, and what other societies must do, is study the experiences of other peoples, and choose policies in the light of the specific needs of cultures, economic systems, and nations. Eventually policies and institutions may be exchanged internationally, just as techniques are now transferred across international borders; but each society must decide for itself whether to follow China's or any other country's mechanization strategy.

What I do hope can be learned from China's experience is that it is possible to impose political will over the path of technological development. Peasants are slowly learning that they are not controlled by inexplicable fate; can we, too, learn that we have the potential of controlling technological society? I hope that this book strengthens our confidence as we try to assert human control over technology.

Research for this book, begun at Columbia University in 1969, has been helped immeasurably by my experience as a research associate at Cornell University from 1972 to 1977. Scores of faculty members and students have shared their insights about rural development throughout the world. Their observations and questions have provided a rich source of stimulation.

I am specially indebted to the China–Japan Program, the Rural Development Committee, the Program on Policies for Science and Technology in Developing Nations, and the Center for International Studies, which have supported me at Cornell during this period. David Mozingo, Norman Uphoff, Milton Esman, John Mellor, David Lewis, Jack Chen, and Yuan-tsung Chen have been constant sources of encouragement and stimulation. Michel Oksenberg has been most helpful, and valuable editorial suggestions have been made by David M. Lampton and Mary Sheridan. Some of the materials for this book were gathered in Hong Kong with the assistance of a grant from the National Science Foundation. The University Service Centre in Hong Kong provided crucial logistical support. The Cornell University Press has provided extensive editorial assistance, for which I am most grateful. Of course, responsibility for errors of fact, judgment, and omission rests solely with me.

BENEDICT STAVIS

Ithaca, New York

Contents

Tables

Figures

THE POLITICS OF AGRICULTURAL MECHANIZATION IN CHINA

1. The Problem of Agricultural Mechanization in China

Of all changes in agricultural technology, mechanization has had the greatest direct social impact. Chemical fertilizers and new seeds may raise yield and income, affect income distribution, and increase the role of technically trained people. Improved irrigation may require new patterns of bureaucratic management and social integration. But mechanization has a far broader social impact. It affects income, income distribution, and the manner in which people relate to their work and to each other. Under certain conditions, agricultural mechanization can increase agricultural production and lighten labor intensity, two indisputably beneficial outcomes. Under other circumstances agricultural mechanization can increase unemployment and cause some farmers to be pushed off the land; great social disruption can result.[1]

Precisely which effect (or combination of effects) agricultural mechanization will have depends on at least three major variables: (1) the type of agricultural mechanization under consideration; (2) the peculiarities of existing and potential cropping systems; this, in turn, depends on the natural environment, available agricultural techniques, the labor supply, and price levels; (3) social institutions, including land tenancy, patterns of urbanization and industrial development, entrepreneurial and managerial skills, and the specific financial arrangements for mechanization.[2]

1. This analysis of mechanization is based on the following: Herman Southworth, ed., *Farm Mechanization in East Asia* (New York: Agricultural Development Council, 1972); Montague Yudelman, Gavan Butler, and Ranadev Banerji, *Technological Change in Agriculture and Employment in Developing Countries* (Paris: OECD, 1972); Carl Gotsch, "Tractor Mechanisation and Rural Development in Pakistan," *International Labour Review* 107:2 (Feb. 1973), 133–166; Frank Child and Hiromitsu Kaneda, "Links to the Green Revolution: A Study of Small-Scale, Agriculturally Related Industry in the Pakistan Punjab," *Economic Development and Cultural Change* 23:2 (Jan. 1975), 249–275; and an unpublished paper by John McInerney and Graham Donaldson on farm tractors in Pakistan.

2. A similar taxonomy for analysis is offered by Carl Gotsch, "Technical Change and the Distribution of Income in Rural Areas," *American Journal of Agricultural Economics* 54:2 (May 1972), 326–341.

These factors interact in extremely complex ways. Even the simple question "Is mechanization profitable?" has many dimensions. The answer depends on many factors:

1. Tractorization of plowing can result in higher yields because better plowing enables plants to develop better root structures.

2. When mechanization replaces animal power, land that had previously been devoted to growing animal food can be put to more profitable use, growing food grains, vegetables, etc.

3. Mechanization of cultivation, harvesting, and processing saves time. In certain regions the time saved may be critical in increasing the cropping intensity. It may permit double-cropping of rice, the development of a wheat-cotton intercropping system, or some other more profitable cropping pattern. Of course, this agricultural intensification requires many other inputs as well, particularly irrigation at appropriate times, fertilizer, and plant varieties that are suited to new time schedules and amounts of sunlight.

4. Mechanization can reduce labor requirements of agriculture. This can bring about an increase in income if alternative, better-paying employment is available, either in specialty agricultural production (animal husbandry, vegetables, etc.) or in the industrial sector. In Japan and Taiwan, agricultural mechanization increases family income by increasing the role of nonfarm income; conversely, off-farm cash income may be needed to purchase machinery.

5. Mechanization of transportation can increase income by giving farmers better access to markets.

6. Mechanization increases "psychic income." In many cases, farmers mechanize because they want to feel part of the modern sector. They want to reduce labor intensity and to use machines. In some cases (as in Japan), they feel that mechanization is necessary to keep young people on the farms. Thus, even if a cost-benefit analysis shows mechanization inappropriate from a narrow financial point of view, the farmers will mechanize. In a sense, machinery is consumer-durable.

7. The profitability of mechanization is, of course, closely related to the cost of mechanization. This is influenced by the price of machines, which is influenced by industrial policies, tariffs, currency controls, etc., and also by the price of energy.

Thus a great many factors influence why and whether agricultural mechanization will be profitable in an overall sense. In some cases, mechanization is subsidized by the state to encourage the use of machines and to benefit the owners and/or users of machines. In other cases, the cost of mechanization is kept artificially high, to extract wealth from the rural sector for the state and/or industrial interests.

Leaving aside the question of profitability, many natural and social factors influence how the benefits of mechanization will be distributed and what the social consequences will be. First, because many items of agricultural machinery come in discrete ("lumpy") sizes which require fairly specific farm sizes to be operated efficiently, mechanization frequently has an effect on farm size and distribution of land. The manner by which farm size is optimized varies from place to place and is closely related to the social structure and land-tenure system. If tenancy is common, very often landlords will purchase tractors and then attempt to cancel tenancy agreements and push the tenants off the land. The landlords monopolize all the benefits of mechanization, and the tenants suffer. This appears to be the basic pattern in Pakistan. If, however, owners cultivate their own land, then some method of sharing machinery can emerge. A group of farmers may pool resources and cooperatively purchase and operate machinery. A farmer may purchase machinery and contract to perform machine services for his neighbors on a custom basis. All these methods have been noted in Taiwan, Japan, and the Indian Punjab, and permit many farmers to benefit from mechanization. Rather little is known about the precise circumstances under which agricultural mechanization will be marked by single farmers expanding their own holdings or by many farmers sharing machines. Undoubtedly the government can play a crucial role either by assisting landlords to evict tenants or by helping to provide the organizational infrastructure for sharing, i.e., helping to organize cooperatives, providing tax incentives for sharing, and the like.

The second distributional effect of agricultural mechanization comes from changes in the labor market. Mechanization can displace farm labor, specifically people who have plowed, harvested, etc. (In certain situations mechanization may also generate demand for farm labor by removing seasonal constraints to increased cropping intensity.) The people who are displaced are often unskilled laborers and have difficulty locating employment. They may suffer greatly from mechanization of agriculture. Of course, mechanization also creates jobs for people who operate, maintain, and otherwise service machines, but these jobs usually require training that is unavailable to the farm laborers who are displaced.

In general, when agriculture is mechanized in countries with much land and limited agricultural labor, the beneficial effects of mechanization predominate. This was true of the midwestern United States. However, where agricultural mechanization has taken place in densely populated regions, many people who were agricultural laborers find themselves unemployed and consider migrating to urban areas in search of industrial employment. If the industrial sector is expanding and can absorb the labor freed by agricultural mechanization (as in Taiwan and Japan), agricultural mechanization seems beneficial to everyone. But if the industrial sector is not expanding, if environmental constraints prevent changes in cropping patterns that would increase the demand for agricultural labor, and if a few people own most of the land while most farmers are tenants, then only a few people—the owners of land and machinery—will benefit from mechanization, and many people will be hurt by it. Tenants and farm laborers will be forced off the land, will find no employment, and may drift to the tenuous life of the slums of a major city. Unfortunately this set of circumstances is tragically common. It is characteristic of many regions in South Asia and Latin America. Thus many people concerned with rural development are arguing that mechanization (especially the use of tractors for plowing) should be discouraged until the employment problems are resolved.

This study will examine how the People's Republic of China has faced the question of agricultural mechanization.[3] China's agricultural mechanization policy has been distinctive in some respects. Generally speaking, China has encouraged farm-tool reform and agricultural mechanization, precisely because mechanization does have so many side effects on social organization. The Chinese leadership has tried, generally successfully, to utilize the side effects of mechanization to accomplish the broader goals of social transformation.

The institutional setting of agricultural mechanization in China has been especially important. The institutions of ownership of both land and machinery have been subject to much debate in China. There are three types of ownership in China: private, collective, and state. With regard to land, private ownership dominated in the early 1950's, but in the mid-

3. Three articles briefly sum up different aspects of China's agricultural mechanization policy: Roland Berger, "The Mechanisation of Chinese Agriculture," *Eastern Horizon* 11:3 (1972), 7–26; Jack Grey, "The Economics of Maoism," *Bulletin of the Atomic Scientists* 25 (Feb. 1969), 42–52; *Current Scene* editor, "The Conflict between Mao Tse-tung and Liu Shao-ch'i over Agricultural Mechanization in Communist China," *Current Scene* 6:17 (Oct. 1, 1968), 1–20.

1950's there was a general change to collective ownership, for reasons which will be examined in detail in Chapter 3. State ownership of land has been limited to a few state farms, generally in reclamation and border areas. Collective and state ownership of land have been crucial in assuring that the benefits of mechanization have been spread broadly within the rural sector. Mechanization has not sharpened cleavages between landlords and tenants, as this distinction has been obviated through collective ownership of land.

The institutional arrangements for owning machinery have also been subject to debate. Since private ownership has been out of the question (except for small farm tools) because individuals do not have enough capital to purchase machines, two patterns of state ownership have been used. First, machinery may be owned by the state and utilized exclusively on state-owned land, i.e., on state farms or state mechanized farms. Alternatively, the state can own the machinery in machine tractor stations (MTS's) and rent out services to peasants, collectively or individually. In this way the state absorbs the initial costs of mechanization and can manipulate the character of the rental contracts to achieve particular policies. For example, mechanical services may be made available only to collective groups of farmers, thus serving as an incentive to pool land. This has been done in China. Second, the costs of mechanical services may be made rather high to extract surpluses from the rural sector and make them available (through the state) for urban-industrial investment. This had been the pattern in the Soviet Union but has not characterized Chinese policy.

The MTS system was important in China in the mid-1950's and again in the early and mid-1960's, but some Chinese leaders felt that it contributed too much to a strengthening of the central state bureaucracy, with potentials for elitism and corruption. Moreover, because ownership of machinery was distinct from ownership of land, there were substantial difficulties in utilizing machinery efficiently and in integrating the manufacture and repair of tools and machines with local industry. Thus, after the Cultural Revolution, the system of state ownership of machines was replaced by collective (commune) ownership of machines. Table 1.1 summarizes the distribution of tractors in different state ownership systems.

Chinese policy toward agricultural mechanization did not emerge overnight. Chinese political leaders started out with very little understanding of agricultural mechanization but with great hopes that it could be used to

Table 1.1. State ownership of tractors in China, 1949–65*

Year	Total tractors in use†	Owned by state farm		Owned by state mechanized farm‡		Owned by machine tractor station	
		No.	%	No.	%	No.	%
1949	401	401	100	–	–	–	–
1950	1,286	1,160	90	1,160	90	30	2
1951	1,410	n.a.	n.a.	n.a.	n.a.	30	2
1952	2,006	1,745	87	1,532	76	30	1
1953	2,719	1,801	66	1,627	60	68	3
1954	5,061	2,766	55	2,235	44	778	15
1955	8,094	4,036	50	2,839	35	2,363	29
1956	19,367	7,243	37	4,422	23	9,862	51
1957	24,629	10,177	41			12,036	49
1958	45,330	16,955	37			10,995	22
1959	59,000	21,000	36			17,300	29
1960	79,000	28,000	35			n.a.	n.a.
1961	90,000	n.a.	n.a.			n.a.	n.a.
1962	103,400	n.a.	n.a.			n.a.	n.a.
1963	115,000	34,000	30			68,040	59
1964	123,000	38,400	32			71,500	58
1965	130,500	40,000	31			79,300	61

Source: Kang Chao, *Agricultural Production in Communist China, 1949–1965* (Madison: University of Wisconsin Press, 1970), pp. 109, 112.

*In this table and all others, unless otherwise noted, numbers of tractors are expressed in 15-h.p. standard tractor units.

†Includes tractors not owned by the state, e.g. owned by communes, research units, universities, etc. Tractors owned by these institutions, by state farms, and by MTS's should total 100 per cent.

‡State mechanized farms were a special category of state farms that existed from 1950 to 1956.

solve many social and economic problems. There was a gradual process of experimenting with policies, getting feedback about the results, considering new policies, struggling politically over alternative policies, and administering new programs.

Because agricultural mechanization policy has been determined through the political process in China, this study will examine the functioning of the political system on this one issue. Several different models exist for ordering data about the Chinese political system. Before the Cultural Revolution the most widely adopted model was the totalitarian model, which emphasized that political leadership was monolithic and drew most of its ideas from Marxist-Leninist doctrine. Shifts in policy were thought to result from resistance of the masses to policies proposed. The Cultural Revolution made clear, however, that the leadership of the

Communist movement in China was by no means unified and monolithic. Following to some extent the Chinese interpretation of recent history, analysts have considered Chinese politics to be a "struggle between two lines," namely the policy views of Mao Tse-tung (revolutionary, socialist, romantic) and those of Liu Shao-ch'i (revisionist, capitalist, pragmatic). More recently, as information became available about Lin Piao and the late phases of the Cultural Revolution, there has been a tendency to see a three-way struggle, between the right wing ("capitalist roaders"), the center, and the "ultra-left." Many interpretations stress the role of Chinese history, the role of Mao Tse-tung as a leader, or the importance of the bureaucracy. A few attempts have been made to interpret Chinese politics in terms of conflicts between factions formed through association in different divisions of the Chinese military, or between interest groups (or their representatives) such as peasants, industrial workers, managers, the military, Party cadres, etc. Other studies hypothesize psychocultural factors, such as conflict-avoidance or factionalism, to explain China's political behavior.[4] The purge of the "gang of four" will undoubtedly provide evidence for all of these interpretations.

This study does not adopt any of these models of Chinese politics. Rather, it will stay very close to the empirical data and will pay particular attention to the choices and constraints imposed on policy by the technology inherent in China's agriculture. The analysis that emerges is of a fairly unified leadership in the early 1950's, somewhat ignorant about the processes and implications of agricultural mechanization. The Soviet model provided the starting point for mechanization policy. By the mid-1950's, some people in China sensed deficiencies in the Soviet model and different perspectives began to crystallize, but there probably was still a great deal of consensus. In addition, the practical constraints were so great that there were few options. By the mid-1960's, the policy options became greater and conflict intensified over mechanization policy.

The debate centered on the social dimensions of agricultural mechanization. Was mechanization simply a way of assuring and increasing the

4. Two excellent works by Michel Oksenberg propose several other models for viewing Chinese politics and offer many examples of scholarship which develop each model: his *Bibliography of Secondary English Language Literature on Contemporary Chinese Politics* (New York: Columbia University East Asian Institute, 1971) and "Political Changes and Their Causes in China, 1949–1972," in William Robson and Bernard Crick, eds., *China in Transition,* special issue of *Political Quarterly* 45:1 (Jan.–March 1974) esp. pp. 107–18.

agricultural productivity of sparsely populated regions? Would it be a way of controlling and taxing the countryside, as in the Soviet Union? Or would mechanization be used as a catalyst for broad social change in the countryside—to consolidate the collective economy and spark rural industrialization?

Unfortunately, for some time the leadership was so concerned with the social consequences of mechanization that in many cases it overlooked the practical technical problems involved. Gradually mechanization policy has become interwined with a broad vision of socioeconomic development. A pattern of industrialization emerged which stressed the role of local industries. These would bring the cultural outlook and productivity of the industrial sector to China's vast rural hinterland. It was a strategy quite different from the prevalent notion that industrialization and modernization will be focused in a few urban areas. The strategy reflects a conscious effort to prevent the development of a dualistic economy, with a wealthy urban sector and an impoverished rural one.

2. Postliberation Policy on Mechanization

During the immediate postliberation years, the main thrust of rural policy was not directly related to mechanization. However, both in rural social organization and in mechanization crucial decisions profoundly influenced the manner in which mechanization was eventually implemented. A radical land-reform program fundamentally altered the rural social and political environment in which mechanization would eventually be carried out. State mechanized farms were established to train people in mechanical skills. Simultaneously a modest effort was made to replace the simple traditional farm tools which had been lost during years of war. After land reform individual farmers were encouraged to form labor exchange groups, known as mutual-aid teams (MAT's). This was the first step in creating a collectivist rural social structure which eventually became the basis for ownership and management of farm machinery. While agricultural mechanization remained negligible in actual amounts during this period, the foundation for future mechanization was being set. These programs will be examined in this chapter.

Land Reform

Land reform included the redistribution of land from wealthy families to families which rented land or worked as landless laborers.[1] Land

1. Excellent studies of land reform in China are John Wong, *Land Reform in The People's Republic of China* (New York: Praeger, 1973); Victor Lippit, *Land Reform and Economic Development in China* (White Plains: International Arts and Sciences Press, 1974); Vivienne Shue, "Transforming China's Peasant Villages: Rural Political and Economic Organization, 1949–56" (Ph.D. diss., Harvard University, 1975); Chao Kuo-chün, *Agrarian Policy of the Chinese Communist Party, 1921–1959* (Bombay: Asia Publishing House, 1960), and *Agrarian Policies of Mainland China: A Documentary Study (1949–1956)* Cambridge: Harvard East Asian Research Center, 1957).

Case studies of land reform in particular localities or regions include William Hinton, *Fanshen* (New York: Monthly Review Press, 1966); C. K. Yang, *Chinese Communist Soci-*

reform served several purposes, not entirely harmonious. First, tenants and laborers would benefit because they would not have to pay rent or be forced to work at minimal wages. This was the most obvious, direct benefit for the rural poor. At the macroeconomic level, the Communist leaders thought that redistribution of land would provide better incentives for higher production and would assure a better standard of living. This required minimizing the disruption of the highly productive rich peasants, so land was to be confiscated from landlords only. Liu Shao-ch'i described this policy very explicitly:

> The basic reason and aim of agrarian reform stems from the demands of production. Hence, every step in agrarian reform should truly take into consideration and be closely co-ordinated with the development of rural production. Precisely because of this basic reason and aim, the Central Committee of the Communist Party of China proposes that rich peasant economy be retained in the future agrarian reform. Rich peasant economy should not be destroyed. This is because the existence of a rich peasant economy and its development within certain limits is advantageous to the development of the people's economy of our country. It is, therefore, also beneficial to the broad peasant masses.[2]

This emphasis on production and this commitment to maintain the rich-peasant economy was not a personal preference of Liu Shao-ch'i (as later charged by Red Guards).[3] It was a deliberate, short-term concession made to rich peasants by a unified party to simplify the earlier stages of consolidating the new regime. Mao Tse-tung had outlined this "rich-peasant policy" even before Liu:

---

ety: The Family and the Village (Cambridge: MIT Press, 1959); Daniel Henry Bays, "Agrarian Reform in Kwangtung, 1950–53," in *Early Communist China: Two Studies* (Ann Arbor: University of Michigan Center for Chinese Studies, 1969); David and Isabel Crook, *Revolution in a Chinese Village: Ten Mile Inn* (London: Routledge and Kegan Paul, 1959).

Chinese fiction offers valuable views of the land-reform process. Two of the best pieces are Chou Li-po's *The Hurricane* (Peking: Foreign Languages Press, 1955), and Ting Ling's "Sun over the Sankan River," *Chinese Literature,* spring 1953, pp. 26–296. An excellent survey of the way in which Chinese Communist literature reflects land reform and many other social movements is Joe C. Huang, *Heroes and Villains in Communist China* (New York: Pica Press, 1973).

2. Liu Shao-ch'i, "On the Agrarian Reform Law," Report at the Second Session of the National Committee of the People's Consultative Conference, Peking, June 1950, available in *People's China* 2:2 (July 16, 1950), 7–8.

3. A Chinese article in 1950 did, however, emphasize the role of Liu Shao-ch'i in this policy: Cheng Lien-tuan, "Why China Preserves the Rich Peasant Economy," *People's China* 2:8 (Oct. 16, 1950), 14.

> Therefore, there should be a change in our policy towards the rich peasants, a change from the policy of requisitioning the surplus land and property of the rich peasants to one of preserving a rich peasant economy in order to further the early restoration of production in the rural areas. This change of policy will also serve to isolate the landlords while protecting the small peasants and those who rent out small plots of land.[4]

Mao's concern about maintaining production had been spelled out in a report the previous year.[5] Another economic reason to preserve the rich peasants was that they could be used as a temporary source for procurement of food for urban areas. Before land reform, urban absentee landlords had collected rent, and this transferred resources from rural to urban areas. After land reform, with rents no longer legal, some other means was needed to feed cities.

While the economic factors of land reform were important, the political factors were dominant. Land reform was the crucial first step in the political transformation of China. Because of the way it was implemented, land reform involved the elimination of the landowning elite as a social, political, and economic force in rural China and the destruction of the existing political structure in rural areas. Preserving the rich-peasant economy served the subtle but important political function of weakening the potential alliance between landlords and rich peasants, thus simplifying the task of isolating, attacking, and destroying the landlord class.

Violence and terror in measured amounts accompanied land reform. Those of the former rural elite responsible for violent crimes against poor peasants (murder, rape, etc.) were sometimes executed. In theory, violence was supposed to be carefully regulated by the government,[6] but

4. Mao Tse-tung, "Report to the Third Plenary Session of the Seventh Central Committee: The Struggle for a Fundamental Turn for the Better in the Financial and Economic Situation of China," (June 6, 1950), *People's China* 2:1 (July 1, 1950), 5. These decisions were made before the outbreak of the Korean War, after which national unity became even more important.

5. Mao said, "Our first tasks are to wage struggles step by step, to clean out the bandits and to oppose the local tyrants (the section of the landlord class in power) in order to complete preparations for the reduction of rent and interest. . . . At the same time care must be taken to maintain the present level of agricultural production as far as possible and to prevent it from declining. In the north . . . the central task of the Party is to mobilize all forces to restore and develop production; this should be the centre of gravity in all work" ("Report to the Second Plenary Session of the Seventh Central Committee of the Communist Party of China," March 5, 1949; available in *Peking Review*, no. 48 [Nov. 29], 1968, 5–6).

6. The Agrarian Reform Law stated: "In the course of agrarian reform, a people's tribunal shall be set up in every county to ensure that reform is carried out. The tribunal shall travel to different places, to try and punish, according to law, the hated despotic elements

Mao Tse-tung himself admitted, "We did not impose much discipline over the settling of that kind of score [i.e., organizing the expression of personal bitterness]."[7] For the whole traditional elite, the show of force demonstrated that the political balance of power had changed and that overt resistance was futile and suicidal. Most members of the former elite became ordinary peasants, although they were "capped" with labels of "landlord" or "rich peasant" and closely supervised to prevent their re-emergence as an elite.

Despite the justifiable emotional heat generated over the question of violence in land reform, remarkably little careful research has been done on the question. An estimate of 14–15 million deaths for 1949–52 has been adopted by some of the standard texts on this period.[8] This estimate was released by the American Federation of Labor's Free Trade Union Committee in October 1952 and published in December 1952.[9] The original report was purported to have been compiled by an underground group in China calling itself "Democratic Revolutionary League," and to have been sent out of China by its secretary Way Min (*Wei Min* meaning "for the people"?) on July 24, 1952. For evidence and sources, the document merely reported "evidence and data obtained by the League; abstracts compiled by the league." No sources were cited which could be checked.

It is highly probable that this report, which has provided the foundation

---

who have committed heinous crimes, whom the masses of the people demand to be brought to justice, and all persons who resist or violate the provisions of the Agrarian Reform Law and other decrees. Indiscriminate arrest, beating or killing of people, corporal punishment and the like are strictly forbidden" (available in Chao, *Agrarian Policies,* p. 43).

7. Quoted by André Malraux, *Anti-memoirs* (New York: Holt, Rinehart, & Winston, 1967), p. 362.

8. Richard Walker, *The Human Cost of Communism in China* (Washington, D.C.: Government Printing Office, 1971), p. 13. This was a report prepared for the U.S. Senate Judiciary Committee, chaired by Senator James Eastland, at the request of Senator Thomas Dodd. The same number is referred to as unspecified "U.S. government estimates," by Franz Michael and George Taylor, *The Far East in the Modern World,* rev. ed. (New York: Holt, Rinehart & Winston, 1964), p. 459.

9. "Mass Murder in Communist China," American Federation of Labor, Free Trade Union Committee *News* 7:12 (Dec. 1952), 1, 4–5. This report was released October 22, 1952, and reported in the *New York Times* Oct. 23, 1952, p. 3. The report offered this breakdown (taken from Table 2): 5 million killed in rural areas (landlords, village despots); 3 million KMT reactionaries; 2.6 million bandit agents; 0.9 million trecherous merchants; 2.1 million died in slave labor camps or in "suicidal missions"; .021 million POW's from Korean War.

for much "scholarship," is bogus.[10] The committee that released the report—the Free Trade Union Committee of the AFL—was funded substantially (if not entirely) by the U.S. Central Intelligence Agency's International Organization Division.[11] One common activity of the CIA is the creation and spreading of propaganda and "disinformation" through a wide range of publications.[12] All the connections cannot be demonstrated conclusively in this case, because the authorship and methodology of the original report remain obscured. In the case of Vietnam, it has been demonstrated convincingly that the CIA helped to finance writers, generate numbers, and spread stories which vastly overestimated the violence of land reform.[13]

A few Chinese sources are helpful in quantifying the extent of violence inherent in land reform. A report from Kwangtung seems to be the most specific. During the ten-month period from October 10, 1950, to August 10, 1951 (the period of greatest tension and violence in Kwangtung's land reform), 28,332 criminals were executed by firing squad (*ch'iang chüeh.*)[14] This figure is the tally of official executions and presumably does not include cases in which peasants illegally (but irrevocably) executed hated landlords. Nor would this figure include suicides of landlords (not infrequent) and deaths suffered in work camps.

10. Attempts to examine the original report from China have been unproductive. According to AFL-CIO officials, the original document (along with all papers of Mathew Woll, chairman of the Free Trade Union Committee) was destroyed. In addition to scholarly books, the *New York Times* accepted the validity of this report completely. After the report was released, *Times* editors intoned: "It is an axiom that life is cheap in crowded and undeveloped countries. It is also an axiom that communist philosophy puts no value on life itself" (*New York Times* editorial, Oct. 24, 1952, p. 22, col. 3).

11. Winslow Peck, "Clandestine Enforcement of U.S. Foreign Labor Policy," *Counterspy* 2:1 (Fall 1974), 36. The close association between the CIA and the AFL in the late 1940's is described by Miles Copeland, *Beyond Cloak and Dagger: Inside the CIA* (New York: Pinnacle, 1974), p. 240.

12. Victor Marchetti and John Marks, *The CIA and the Cult of Intelligence* (New York: Dell, 1974), pp. 165–185.

13. D. Gareth Porter, "The Myth of the Bloodbath: North Vietnam's Land Reform Reconsidered," Cornell University International Relations of East Asia Project, Interim Report 2, 1972; summarized in *Indochina Chronicle,* no. 19 (Sept. 15, 1972), 1–5. Funded by the CIA (using the Congress of Cultural Freedom as a conduit) and USIA, Hoang Van Chi wrote that up to 700,000 people (5 per cent of the population) were killed. He offered no sources, and mistranslated Vietnamese documents to justify these figures. These estimates were then picked up by others writing on Vietnam (including President Richard Nixon). Porter presents careful documentation for an estimate that 800 to 2,500 people may have been killed in land reform—only 0.1 to 0.4 per cent of the widely adopted numbers.

14. Ku Ta-ts'un, "Report on the Work of the Kwangtung Provincial Government during the Past Ten Months," *Nan-fang Jih-pao,* Sept. 18, 1951 (CB 124 p. 4).

A crude extrapolation of the Kwangtung figure is illuminating. Presume that as many people were killed "unofficially" as officially. If 60,000 were killed altogether in Kwangtung, this represents 0.17 per cent of the population of 35 million.[15] If this is extrapolated on the basis of a population for all of China of 500 million, then it implies that roughly 850,000 people may have been killed. Of these, landlords killed in the process of land reform represented a small portion.

Another regional figure is available, but it has been used erroneously. Richard Walker reports that Teng Tzu-hui claimed that 322,000 people had been executed in the Central-South Region from the time of liberation to November 1951.[16] Teng does say that 1,150,000 native bandits (*t'u fei*) were inactivated (*hsiao mieh*) and that of the serious criminals (*tsui-ta e-chi te jen-fan*) 28 per cent were executed (*ch'u szu*).[17] Walker has assumed that local bandits and serious criminals are the same category, but this seems unlikely. Probably nothing can be said about the absolute amount of violence in land reforms from this speech of Teng Tzu-hui.

Perhaps the best figures come from Mao Tse-tung and Chou En-lai. In 1956, Mao said, "In the past, we have killed [*sha*], locked up [*kuan*], and controlled [*kuan*] 2–3 million [counterrevolutionaries], and this was extremely necessary. Without this stroke, it would not do."[18] Of these 2–3 million, how many were killed? Chou En-lai reported that "16.8 per cent of the counter-revolutionaries dealt with were sentenced to death because they had committed heinous crimes and public wrath was strong against them. The great bulk of these sentences were passed between the time of liberation and 1952. This was absolutely necessary at the

15. When Chinese Nationalist troops "pacified" Taiwan Province in March 1947, about 10,000 people (mostly students, intellectuals, independent businessmen, and politicians) were killed. This represents 0.15 per cent of Taiwan's 6.5 million inhabitants at the time. The estimate of 10,000 deaths is by George Kerr, "Formosa: The March Massacres," *Far Eastern Survey* 16 (Nov. 5, 1947), 225. The incident is described in graphic detail in *United States Relations with China, with Special Reference to the Period 1944–49* (Washington, D.C.: Department of State Publication 3573, Far Eastern Series 30, 1949; reprinted by Stanford University Press, 1967), pp. 923–938.

16. Richard Walker, *China under Communism* (New Haven: Yale University Press, 1955), p. 218.

17. Teng Tzu-hui, "Report on the Work of the Central-South Military and Administrative Commission," *Ch'ang Chiang Jih-pao,* Dec. 13, 1951 (CB 157, pp. 11, 12).

18. Mao's Speech at Expanded Meeting of CCP Political Bureau, April 1956, in *Miscellany of Mao Tse-tung Thought* (Arlington, Va.: Joint Publications Research Service, 1974), p. 34. Hereafter this volume will be referred to as *Miscellany*.

time."[19] This would imply that somewhere between 336,000 and 504,000 people were sentenced to death, presumably after the establishment of the new government in 1949. (Not all of those sentenced were actually executed, although there is no way of knowing for how many people execution was indefinitely delayed.)

This figure is reinforced by other bits of data. In his speech "On the Correct Handling of Contradictions among the People," of February 27, 1957, Mao is reported to have specified the number of counterrevolutionaries who had been liquidated. According to one person, the number was 500,000;[20] according to a report which got to the *New York Times* (from the Central Intelligence Agency?) the number was 800,000.[21] (The numbers were deleted from the published version and official translation of this speech.)

Yet one other bit of data should be considered. Chou En-lai is reported by Edgar Snow to have said (whether publicly or privately is not specified) that 830,000 "enemies of the people" had been destroyed (*hsiao-mieh*) during the war over land confiscation, mass trials of landlords, and subsequent supression of counterrevolutionaries.[22]

In these circumstances it is impossible to know how many people were officially executed in land reform, how many were lynched by enraged peasants, how many committed suicide. Rough orders of magnitude can, however, be suggested. It would appear that somewhere between 400,000 and 800,000 people were killed officially after 1949. What portion of these were landlords and members of the rural power structure (including rural police, Nationalist Party (Kuomintang) troops and commanders, rent collectors, etc.) cannot be said for certain, but it is possible that the rural revolution could have cost 200,000 to 800,000 lives. The Chinese Communist leadership had estimated that landlords and their families constituted 4–5 per cent of the rural population—about 20 million people.[23] This would imply that 1 to 4 per cent of landlords' families met death. If a half-million people were killed in land reform, this would be 0.1 per

19. Chou En-lai, "Report on the Work of the Government," Fourth Session of the First National People's Congress, June 26, 1957; available in Supplement to *People's China*, no. 14 (July 16), 1957, p. 7.

20. Mu Fu-sheng, *The Wilting of the Hundred Flowers* (New York: Praeger, 1962), p. 128.

21. *New York Times*, June 13, 1957, p. 8.

22. Edgar Snow, *Red China Today* (New York: Vintage, 1971), p. 346.

23. Chao, *Agrarian Policies*, pp. 34, 36.

cent of the rural population or 2.5 per cent of the landlord class and would represent roughly one death in six landlord families.[24]

This represents substantial violence, but far less than commonly believed. Moreover, the violence and loss of life inherent in revolutionary land reform should be compared with the violence and loss of life in the "normal" situation. There are no statistics available on how many people each year before land reform were beaten to death for failure to pay rents.[25] A very crude point for comparison is the number of people killed in natural disasters in the years before land reform. These data are summarized in Table 2.1 and show that "normal" natural calamities involved as much loss of life as land reform.

In redistributing land, the Chinese took care to clarify class relations. Households were not simply classified according to how much land they owned. Rather, they were classified in theory according to the percentage of their income which was based on exploiting other people, through rents, interest, or hired labor. Of course, in practice the classification was very difficulty. Could landlords be classified instead as merchants or small industrialists if they engaged in these activities? What about wealthy peasants who supported revolution for reasons of nationalism? How would class categories be established where villagers perceived cleavages in terms of lineages and villages?[26] In such situations class lines were not at all clear, and there was room for negotiation, personal grudges, and quotas in demarkation of classes. Nevertheless, land reform

24. One village for which detailed information on violence is available is Long Bow, the village described by William Hinton in *Fanshen*. In this village of about 1,000 residents, about twelve people were killed or committed suicide, including local police, landlords, rich peasants, and religious leaders. This represents over 1 per cent of the population. About half the landlords in the village were killed; the other half ran away. This level of violence was atypically high. In Long Bow, land reform came in the middle of civil war and was not moderated by cross-cutting family or clan ties. Initial leadership was weak and Communist Party supervision was lacking. Hinton was able to collect data for *Fanshen* precisely because higher Party levels sent an investigation team to find out why there had been such unusual amounts of violence.

25. In Long Bow, at least seven people starved to death in the famine of 1942 when wealthy refused to share food with less fortunate; at least two tenant farmers had been beaten to death by landlords for minor transgressions of local mores (Hinton, *Fanshen,* pp. 51–52).

26. It is interesting to contrast the situation in Long Bow, with its deep class cleavages unmitigated by loyalties to lineages, with the situation in rural Kwantung, where class conflict was much less. See Ezra Vogel, *Canton under Communism* (Cambridge: Harvard University Press, 1969), pp. 102–106.

Table 2.1. People affected by disasters in China, 1878–1954

Year	Region	People affected (millions)	People killed (millions)
1878[a]	North China Plain drought	50	10
1887[b,c]	Yellow River flood, Honan		1–7
1911[b,d]	Huai River flood	2.7	"immense loss"
1917[c]	Hai River flood, Tientsin (12,700 sq. mi.)	5.6	
1920–21[b]	North China drought	20	.5
1921[e]	Yellow River flood, Shantung	.25	
1928–30[f]	Northwest China drought (Kansu, Shensi, Suiyuan)	10	3–6
1931[g]	Yangtze River flood (84,000 sq. mi.)	9	
	North China floods (84,000 sq. mi.)	11	
	Huai River flood (84,000 sq. mi.)	19	
	Nationwide	40–50	
1933[h]	Yellow River flood	3.6	.02
1935[i]	Han River flood, Hupeh	3.7	.08
1938[h,j]	Yellow River dikes blown up, Honan	12.5	.89
1954[k]	Yangtze, Huai Rivers flood (42,000 sq. mi.)	10	"few"

Sources:

a. Paul Richard Bohr, *Famine in China and the Missionary* (Cambridge: Harvard University East Asian Research Center, 1972), p. 26.

b. Walter Mallory, *China: Land of Famine* (New York: American Geographical Society, 1926), pp. 2, 49–53.

c. John R. Freeman, "Flood Problems in China," *Transactions of the American Society of Civil Engineers* 85 (1922), 1405.

d. "Famine Notes," *Chinese Recorder* 43 (1912), 91; Wm. F. Junkin, "Famine Conditions in North Anhui and North Kiangsu," *ibid.*, p. 75.

e. Oliver Todd, "Taming 'Flood Dragons' along China's Huang Ho," *National Geographic* 81:2 (Feb. 1942), 205–234.

f. Edgar Snow, *Red Star over China* (New York: Grove, 1968), pp. 214–218; G. Findlay Andrew, "On the Trail of Death in Northwest China," *Asia* 32:1 (Jan. 1932), 42–48, 60–62.

g. "Flood Damage in China during 1931," *Chinese Economic Journal* 10 (1932), 341–352; Edmund Clubb, "Floods of China, a National Disaster," *Journal of Geography* 31:5 (May 1932), 199–206; George Stroebe, "The Great Central China Flood of 1931," *Chinese Recorder* 63 (1932), 557–680.

h. *China Tames Her Rivers* (Peking: Foreign Languages Press, 1972), p. 5.

i. "Harnessing the Han River—Major Yangtze Tributary," NCNA Peking, Feb. 5, 1976.

j. O. J. Todd, "The Yellow River Reharnessed," *Geographical Review* 39:1 (Jan. 1949), 38–56.

k. Hsieh Chueh-tsai, report to the National People's Congress, Sept. 26, 1954, NCA Peking, Sept. 29, 1954.

popularized the idea that the manner in which people related to means of production and the character of income were just as important as how much land was owned and the quantity of income.[27]

Another aspect of land reform was psychological. Land reform was implemented in a manner to encourage mass participation in the process of shaping village relationships by poor peasants and women and other formerly powerless people. To generate participation and a feeling of power among traditionally exploited people, passive for centuries for reasons of self-defense, required a psychological revolution far more than economic redistribution.[28] In the long run this cultural transformation may be the most important dimension of land reform.

One outcome of land reform, of course was that the Communist Party became dominant politically in rural China. It had organized land reform, inspired the labeling of people according to different social classes, stirred up mass participation, and actually distributed land. It had extensive military backing, both locally through people's militias and nationally through the People's Liberation Army. The shift in rural political power from the previous elite to the Communist Party was not irreversible; many activities throughout the years after land reform were needed to maintain the new balance of political power. Moreover, the control of the Communist Party was not complete. With their new power of confidence, peasants demanded sensitive, suitable policies of their new leaders. For the Party to maintain a position of leadership, it had to meet demands of the formerly poor peasants.

The elimination of the traditional rural elite as a dominant social class made it possible to carry out the collectivization of agriculture in a peaceful manner, and in a broad sense established the overall institutional framework in which mechanization took place. Land reform meant that the traditional elite could not monopolize the benefits of mechanization.

State Mechanized Farms

During this period of land reform, agricultural mechanization was a low priority. China had just a few hundred tractors. In the early 1950's there really was no viable alternative to direct state ownership of tractors.

27. Philip C. C. Huang, "Analyzing the Twentieth-Century Countryside: Revolutionaries versus Western Scholarship," *Modern China* 1:2 (April 1975), 147–50.

28. Hinton, *Fanshen,* p. vii.

Land reform tore apart the class of rich capitalist-farmers, the farmers with sufficient capital to purchase tractors, with sufficient land to use tractors efficiently, with sufficient managerial ability to operate and maintain modern equipment. Moreover, there was no infrastructure for supplying fuel, parts, or repairs. Before that time, tractors were all owned and managed by some central government authority, such as state farms operated by Japanese in wartime occupied regions or the United Nations Relief and Rehabilitation Administration after the war.

In the first years after liberation, the state mechanized-farm system (data for which are offered in Table 2.2) predominated for several rea-

Table 2.2. State mechanized farms, 1950–56

Year	No. of farms[a]	Total standard tractors[b]	Average tractors/ farm[c]	Total area (million ha.)[a]	Area/ farm (ha.)[c]	Area/ tractor (ha.)[c]
1950	36	1,160	32.2	.09	2,480	77
1952	50	1,532	30.8	.14	2,707	88
1953	59	1,627	27.6	.14	2,393	87
1954	97	2,235	23.0	.19	1,907	83
1955	106	2,839	26.7	.27	2,540	95
1956	166	4,422	26.6			

Sources:

a. Kang Chao, *Agricultural Production in Communist China, 1949–1965* (Madison: University of Wisconsin Press, 1970), p. 112.

b. *Ibid.*, p. 109.

c. Computed.

sons. It was, simply put, convenient. It represented continuity with the previous forms of quasi-state ownership. Second, state mechanized farms served a particular role in ideology. According to Marxist theory, especially as developed by Stalin in the Soviet Union, state farms are considered more "advanced," because a greater portion of the property is owned by the state, representing all of the people. In contrast, the MTS system leaves ownership of land and agricultural produce in the hands of the farmers, individually or collectively. The ideological pressures have been weak however. Despite Marxist theory, China has never moved in the direction of making state farms the primary pattern of rural social-economic organization. Periodic theoretical statements suggest that the state farm is the appropriate form of organization in the distant future, but no steps have been taken to make this a trend.[29] Major problems, to be discussed below, have stood in the way of expanding the state farm system.

State mechanized farms developed in the early 1950's (and have been maintained) for limited practical reasons. First, state mechanized farms could supply commodity grain for urban consumption from previously uncultivated land.[30] Some farms could guarantee production of certain commercial crops, such as cotton.

Second, state mechanized farms could have important military and political implications when they open virgin lands in sparsely populated areas that are near international borders or are inhabited by minority peoples. The strategic role is highlighted by the fact that mechanized state farms are often used to provide employment for army veterans. Some of the farms in Sinkiang and in the Northeast in particular seem to have strong military overtones and were established by veterans.[31] The military-strategic role of state farms was particularly important after the early 1950's when the Soviet Union was no longer a close ally. In 1965, T'an Chen-lin, a politbureau member and specialist in agriculture, pointed out the military importance of opening up these sparsely popu-

29. At one point Mao indicated that state farms would grow in relative importance, but he did not indicate that they would become in the foreseeable future the dominant form of social organization. Mao Tse-tung, Summing-up Speech at Sixth Expanded Plenum of the Seventh CCP Central Committee, Sept. 1955, *Miscellany,* p. 24.

An important theoretical statement in early 1975 recommitted China to the ultimate goal of expanding state farms, but also put this transition into the cloudy future: "Ownership by the whole people is as yet rather weak in agriculture, the foundation of the national economy. The non-existence of bourgeois right in the realm of the system of ownership in a socialist society, as conceived by Marx and Lenin, implies the conversion of all the means of production into the common property of the whole society. Clearly we have not yet advanced to that stage. Neither in theory nor in practice should we overlook the very arduous tasks that lie ahead of the dictatorship of the proletariat in this respect" (Chang Chun-chiao, "On Exercising All-Round Dictatorship Over the Bourgeoisie," *Peking Review,* no. 14 [April 4], 1975, p. 7).

30. This would appear to be the case with many of the farms in the Northeast, in Honan in the region flooded by the Yellow River ("Winter Wheat Yields in State Farms Increased," NCNA Peking, July 20, 1953; SCMP 616, p. 19), and in Hopei by the Po Hai, where the soil was saline and alkaline ("Over 50 Mechanized State Farms in China," China News Service, Peking, Feb. 25, 1954; SCMP 763, p. 18).

31. "Modern Mechanized Farms Started by PLA in Sinkiang," NCNA Tihua, March 28, 1952 (SCMP 306, p. 28); "Two New Mechanized Farms to be Set up in Sinkiang," NCNA Tihua, Jan. 2, 1954 (SCMP 719, p. 41); "Ten Big State Farms in Northeast Draw up 1954 Plan to Deliver 40,914 Tons of Grain to the State," NCNA Mukden, Feb. 26, 1954 (SCMP 758, p. 20); "State Owned Farm 852 Produces Remarkable Results in the Five Years after Its Establishment," *Jen-min Jih-pao,* Nov. 7, 1961 (SCMP 2623, p. 21).

lated areas: "We have so many mountains that we shall hardly know if secret agents are dropped [from the air] . . . so we have to occupy the mountain strongholds."[32]

Third, mechanized state farms could be used as sites to test machines and to train personnel in operation, maintenance, repair, and management.[33] Chiheng State Mechanized Farm in Hopei, where William Hinton worked, was primarily a school in mechanized agriculture.[34] Later, training of personnel was a major task at Sino-Soviet Friendship Farm, sponsored by the Soviet Union near the Soviet border with Heilungkiang in 1954.[35] Personnel trained at these state mechanized farms became crucial technical personnel in later years.

Despite these practical advantages, there were also serious problems with the state mechanized farm system which prevented the widespread adoption of this form of organization. The most obvious problem was that the state farm system was not adapted technologically or organizationally to deal with the problems of settled, intense agriculture which characterized most regions of China. The state farms had the machinery, skills, and organization to break open virgin land, to carve out fields several kilometers long and wide, and to utilize large, sophisticated, expensive machinery. This technology and social organization, imported from the Soviet Union, may have made sense on the Russian steppes and in some areas of China, but it simply was unsuitable for most parts of China in the early 1950's. There was no way that individual farmers after land reform or even the mutual-aid teams and cooperatives which were eventually established could afford to buy such large machines. Nor did they have the skills to maintain or utilize them. Nor were their fields or cultivation systems suitable for large-scale machinery. In short, the mechanical technol-

32. "T'an Chen-lin's Speeches on Resettlement Work," *Chi-nung Hung-ch'i* [Support the Peasant Red Flag], Canton, no. 7, Jan. 1968 (SCMP 4123, p. 12). Perhaps because of the military connections, it was convenient in at least one case to use state farms as prison labor camps. See Bao Ruo-wang and Rudolph Chelminski, *Prisoners of Mao* (New York: Penguin, 1976), p. 12.

33. "State Farms in Kwangtung Plant Subtropical Crops," NCNA Canton, June 10, 1954 (SCMP 828, p. 30).

34. William Hinton, *Iron Oxen* (New York: Vintage, 1971), pp. 149–150. Twenty-five years later some of Hinton's former students were working in farm mechanization near Peking. "William Hinton Tells about Progress in Long Bow and a Visit to Tsinghua," *China and U.S.* 5:4 (July–Aug. 1976), 16.

35. Audrey Donnithorne, *China's Economic System* (New York: Praeger, 1967), p. 92, n. 3; Chang Liu-ch'ih and Wang Cheng-lin, "The State Operated Yu-Yi (Friendship) Farm—A Classic Example of Mechanized Farm and School," *Jen-min Jih-pao,* Nov. 12, 1959 (SCMP 2155, p. 14).

ogy and social organization represented by the state farm system could not be applied to most of the populated regions of China.

In theory, the state mechanized farms were supposed to assist in popularizing modern agricultural techniques among peasants in the surrounding countryside. In general, however, peasants realized that they could not procure the modern inputs (such as machinery and fertilizer) that were available to the state mechanized farms, so the extension function was rarely performed. At the same time, there was a tendency for state farm personnel to ignore the extension work because dealing with traditional technology was beneath their dignity. Indeed the state mechanized farms were quite isolated from the surrounding countryside.[36]

Even in regions for which they were technically and organizationally adapted, the state farms had serious management problems which often resulted in economic loss. Management problems stemmed from the fact that with a few exceptions, the mechanized state farms were established in barren, unpopulated areas by people with little farming experience, including many demobilized soldiers. Profitable operation was difficult. At each of the yearly conferences held to review their work, the farms were criticized for inefficiency and were urged to tighten up management and accounting procedures. The similarity of the message each year suggests that improvement was negligible. An article in June 1953 indicated the broad outlines of the problem:

> Not a few state farms have been detached from actual conditions . . . in respect of their operation and management, agricultural mechanized technique, exemplary functions, and relationship with the masses.
>
> . . . One-sided strife for larger output has arisen, the lowering of cost has been ignored, and the deviation towards a bread-and-butter attitude has seriously come into existence.
>
> [In three state farms] each farm management has blindly applied more fertilizer, carried out more irrigation, engaged more workers and spent more money in order to accomplish the task of planned production. But as a result over-application of fertilizer, bad irrigation, and improper technical measures have brought about a decrease of output.[37]

36. Interview with Chinese tractor technician.

37. "Some Concrete Problems in the Work of State Farms," *China Agricultural Fortnight,* June 25, 1953 (SCMP 634, p. 13). In other years similar problems were expressed: "Summary Report on the Conference on North China State Farm Work, by Department of Rural Work of the North China Bureau of the CC, CCP," *China Agricultural Journal,* no. 4 (Feb. 25), 1954 (SCMP 783, supplement p. i); "Meeting of Directors of Large-Scale Farms in East China Summarizes Experiences, Preliminary Determination of Plans for 1952." *Chieh-fang Jih-pao,* Dec. 31, 1951 (SCMP 271, p. 29); N.E. Bureau, CC, CCP,

The review of the work of state farms for 1957, given by Wang Chen, Minister of State Farms and Land Reclamation, in early 1958, indicated that these problems had not yet been solved.

> Administration and management still show many serious drawbacks. . . . Unnecessary spending of money still reaches a serious proportion. The labor power of some farms is not well organized and labor productivity is very low. The rated capacity of machines is utilized to a low extent. The average wages for agricultural and pastoral workers are too high, and extra wages are paid indiscriminately according to the system of factories and mines (extra wages amount to 12.5 per cent of basic wages). Nobody assumes responsibility for safekeeping and use of supplies.[38]

Wang Chen's report in early 1960 sounded like those from previous years.[39]

For the 1960s, from the little information available, it appears that the trends established during the 1950's continued. Tractors on state farms continued to increase, but the relative importance continued to decline. By 1964, state farms owned 32 per cent of China's tractors.[40] This would amount to roughly 38,400, or approximately a 37 per cent increase over 1960. It was reported that management was improved. More careful economic analyses of agricultural techniques were carried out.[41] Moreover, a system of material incentives was introduced by establishing work points and smaller production teams, at least in some state farms, so that wages would be related to both individual and small-group effort.[42] During the Cultural Revolution, this form of management was criticized, and was claimed to have been supported by T'an Chen-lin, politbureau agricultural specialist.

To summarize, since the early and mid-1950's, state mechanized farms have played an important part in China's mechanization of agriculture.

"Decisions on Strengthening Work of State Farms, September 1952," *Jen-min Jih-pao*, Dec. 10, 1952 (SCMP 485, p. 16).

38. Wang Chen, "Exert Revolution Efforts to Achieve a Rapid Advance in State Farm and Ranch Production" (Speech to First National Congress of Chinese Agricultural and Water Conservancy Workers' Trade Unions, Jan. 30, 1958), Peking, *Kung-jen Jih-pao*, Feb. 1, 1958 (SCMP 1723, p. 2).

39. Wang Chen, "Strengthening the Construction of State Farms," *Hung-ch'i*, no. 7 (April 1), 1961 (SCMM 258, p. 6).

40. Shang Chi-lung and Ma Ching-p'o, "Fifteen Years of Agricultural Mechanization on State Farms," *Nung-yeh Chi-hsieh Chi-shu*, no. 11 (Nov.), 1964 (SCMM 451, p. 6).

41. Ting Li-ch'u, "Exploration into the Rational Management of the State Farm: Investigation and Study of the Problem of the Management of State Farms in Heilungkiang," *Ching-chi Yen-chiu*, no. 12, 1963 (SCMM 403, p. 12).

42. Shang and Ma, "Fifteen Years," SCMM 451, p. 6.

They have, however, been used primarily to train personnel and to achieve a strategic presence in sparsely populated regions of inner Asia and Northeast China. (In 1975, in Heilungkiang, 125 state farms had roughly one-third of the farmland.)[43] There has never been an effort to make state farms the primary institution for owning and managing agricultural machinery.

Supplying Small Farm Tools

After land reform, in addition to developing state mechanized farms, the government gave attention to supplying traditional farm tools and improved farm tools. This seems to have been done in a manner to encourage pooling of agricultural resources in preparation for the formation of agricultural cooperatives.

When the Communist leadership assumed power in 1949, there was a massive tool shortage. China had lost perhaps 30 per cent of its farm tools during the years of international and civil war.[44] An UNRRA report described the problem in graphic terms: "One result of the Japanese occupation was the confiscation of metal of all kinds for war uses. In many rural areas this resulted in a serious shortage of farm implements and hand tools. Many tools were also abandoned by farmers, fleeing ahead of the Japanese advance. The normal replacement of farm tools was prevented since blacksmithing and other farm tool production facilities were lost."[45]

The Chinese responded to the shortage in early 1951, when the Ministry of Agriculture held its first National Conference on Farm Implements. Plans were made for new factories and repair depots to supply and repair traditional farm implements. The plan called for the production of

43. FBIS, May 10, 1976, L3.

44. *Ching-chi Yen-chiu,* no. 3, 1957, pp. 100–126, cited by Chi I-chai, "A Study on the Production and Need of Agricultural Machinery on China Mainland," *Fei-ch'ing Yüeh-pao,* T'aipei, 10:3 (May 31, 1967), 63–72 (JPRS 42, 271, p. 2); *Hsin-hua Yüeh-pao,* Nov. 1952, p. 162, cited in *ibid.;* Li Ching-yü, "Realize Step-by-Step Agricultural Mechanization through Tool Innovation," *Chung-kuo Nung-pao,* Oct. 8, 1959 (ECMM 194, p. 33); Eighth Ministry of Machine Building, Mass Criticism Unit of Revolutionary Great Alliance Committee, Agricultural Machinery Management Bureau, "History of Struggle between the Two Lines (on China's Farm Machinery Front)," *Nung-yeh Chi-hsieh Chi-shu,* no. 9, 1968 (SCMM 633, p. 1); "Problems of Spring Cultivation in Ningpo Area," *Chieh-fang Jih-pao,* March 5, 1952 (SCMP 306, p. 32).

45. United Nations Relief and Rehabilitation Administration, *Agricultural Rehabilitation in China,* Operational Analysis Papers, no. 52 (Washington, D.C.: UNRRA, April 1948), p. 34.

about 10 million implements such as plows and harrows during 1951.[46] By 1954 production reached about 58 million. In addition, China imported a large number of rubber-tired wheels to improve farm carts.[47]

A crude way of evaluating the adequacy of this supply is to estimate what would have been required to cover the requirements of population growth and depreciation. If we assume that 4 per cent of the existing tool stock must be replaced each year due to depreciation (actually, most tools have a life span of less than twenty-five years)[48] and add 2 per cent to cover population growth, then 6 per cent of the existing tool supply must be produced each year. Simply to maintain the 721 million tools from the 1930's[49] would therefore require the manufacture of 43 million tools. Not until 1953 or 1954 did China begin to make this number of tools. Of course, this supply would simply maintain the existing ratio of tools per family. It could not recover previous losses. Thus not until 1954, at the earliest, could the gap with the prewar tool supply begin to be closed.

It is no wonder that the Kwangtung Provincial Land Reform Committee discovered in February 1953 that the "prices of agricultural tools are higher than before the liberation." The price of a plow had risen from eighteen catties or grain to forty catties. (Two catties equals one kilogram). The committee urged an increase in output and supply of agricultural tools.[50]

After land reform the main social policy in the countryside was the development of mutual-aid teams (MAT's), in which peasants exchanged labor to help each other at times of peak labor needs. While collectivist ideological pressure may have been one imporant reason for this mass movement, the tool shortage was another reason. The MAT's facilitated sharing in the use and purchase of farms tools. The shared use of tools, in this period of great shortage, was more efficient.

46. "More and Better Farm Implements to be Provided by State," NCNA Peking, Feb. 27, 1951 (SCMP 74, p. 15).

47. The contracts were for 900,000 and had to be divided among three manufacturers to meet delivery schedules. Roland Berger, "The Mechanisation of Chinese Agriculture," *Eastern Horizon* 11:3 (1972), 8.

48. Pan Hong-sheng and O. T. King, "Preliminary Note on an Economic Study of Farm Implements," *Economic Facts* (University of Nanking, College of Agriculture and Forestry), no. 3 (Nov. 1936), 155.

49. Computed on the basis of data from John Buck, *Land Utilization in China Statistics* (Nanking: University of Nanking, 1937), pp. 394–399.

50. Investigation Section, Kwangtung Provincial Land Reform Committee, "Mass Reaction and Problems after Land Reform," *Nan-fang Jih-pao,* Feb. 19, 1953 (SCMP 527, p. 21).

Even before the postwar tool shortage, the economic logic of sharing tools was clear. Data from the 1930's (offered in Table 2.3) show that although the larger landowners possessed more tools, they had fewer tools per unit of land. This indicates underutilization of tools by the small land-

Table 2.3. Tools per hectare on Chinese farms, 1929–33

Size of farm*	Ha.[a]	No. of tools[b]	Tools/ha.
Small	.97	9.9	10.1
Medium	2.47	16.3	6.7
Large	5.75	27.4	4.7

Sources:
a. John Buck, *Land Utilization in China* (Nanking: University of Nanking, 1937), p. 438.
b. John Buck, *Land Utilization in China, Statistics* (Chicago: University of Chicago Press, 1937), pp. 394–399.
*The "size of farm" refers to the "Standard of Living Survey," not the "Farm Survey," in the sources cited.

owners, who had to own more tools than necessary because of lack of sharing. If land could have been pooled in some way, the tools could have been used more efficiently.

William Hinton reported this to be precisely the case in south Shansi;

> Primitive as the means of production were, no family could afford to have a complete set of the implements and draft animals necessary for production. One good mule could farm 20 acres [8 hectares]. But no family owned as much as 20 acres, or even half that much. The largest holding in all of Long Bow was a little more than eight acres for a family of nine people. Even this large family could not use all the productive capacity of one mule. If they had tried to keep a mule and a cart and a plow and a harrow and a seeder, the capital investment would have been grossly out of proportion to the production possible on eight acres of land. Therefore, a family that owned a mule was not likely to own the other essential implements, not even a cart, even though the latter was used as much for transport on the roads as it was for agriculture.[51]

Investigators from the University of Nanking's School of Agriculture and Forestry, surveying in East China in 1935, noticed the same problem: "The use of work animals and larger implements is limited by size of farm as well as by inconveniently small fields. Both these limitations might be overcome if farmers would cooperate in using work animals and

51. Hinton, *Fanshen*, p. 212.

larger implements, in rearranging the layout and size of their fields, and in levelling graves.''[52]

There were traditional mechanisms for sharing tools in the 1930's. On sixty farms in Suhsien, Anhwei, a third or more of the farmers borrowed plows, seed drills, and carts. A quarter borrowed harrows and stone rollers and drags.[53]

Mutual-aid teams were established in 1943 in the Shen-Kan-Ning liberated zone, and a conscious effort was made to utilize the existing pattern of labor sharing. An editorial in *Liberation Daily* proposed:

> In the villages of the border regions, there have traditionally existed all ways for pulling together a labor force, such as *pienkung* and *chakung,* which are methods rather widely used. These voluntary *pienkung* and *chakung,* though rather narrow in scope, and only restricted to relatives, friends, neighbors, are nevertheless suited to the concrete conditions of villages in the border regions. If they can be effectively utilized and directed, and organized and led in a planned way, then they can be transformed into organizations for developing productivity and raising production.[54]

MAT's were able to utilize farm tools as well as labor more efficiently. In addition, they were better able to finance improved farm tools—such as plows—that were too expensive for an individual farmer.[55] It seems significant that there was a sharp increase in the production of improved tools (particularly walking plows and water wheels) in 1950–52, when MAT's were being formed. From 1952 to 1955, however, the number of these tools introduced leveled off or declined slightly; this coincided with a drop in growth of the MAT's. The supply of plows, waterwheels, and other tools could have had a modest social effect. In 1952, less than one million tools were supplied, and about 50 million families were involved in roughly 5 million MAT's or other forms of agricultural cooperation. Thus up to one-fifth of the MAT's could have purchased an improved tool by 1952, and the other MAT's had reasonable expectations of buying some new, improved tools in the near future.

52. Pan and King, ''Preliminary Note,'' p. 157.

53. Pan Hong-sheng and John Raeburn, ''Ownership and Costs of Farm Implements and Work Animals in Suhsien, Anhwei,'' *Economic Facts,* no. 7 (Oct. 1937), 300.

54. ''Let's Organize the Labor Force,'' *Chieh-fang Jih-pao* editorial, Jan. 25, 1943; quoted in Franz Schurmann, *Ideology and Organization in Communist China* (Berkeley: University of California Press, 1968), p. 421. The traditional origin of the MAT's has been clearly shown by John Wong in ''Peasant Economic Behavior: The Case of Traditional Agricultural Co-operation in China,'' *Developing Economies* 9 (1971), 332–349.

55. An agrotechnician formerly in China stressed this aspect in an interview.

3. Mechanization and Consolidating the Collective Economy

In most countries agricultural mechanization begins when a few farmers buy tractors. Experience develops slowly and in a scattered manner. Finally a broad policy may be adopted. In China, the experience was precisely the reverse. From the beginning, broad theoretical questions of state policy shaped agricultural mechanization policy long before much empirical experience had been gathered. While this assured that agricultural mechanization would fit overall state plans and objectives, it also meant that policy would not be based on the concrete conditions of China's agronomic realities. This chapter examines how the broad question of collectivization of agriculture shaped agricultural mechanization programs in the early and mid-1950's. The next chapter shows how agricultural mechanization policies were conditioned by broad issues of state institutional structure.

In the Soviet Union agricultural mechanization had been a crucial element in broad state policy. Tractorization helped the consolidation of the collective economy, simplified state procurement of food, and strengthened political control over the rural sector. Would the same be true in China? During the first years of the 1950's, there was widespread consensus in China's leadership that the Soviet model, in rough outline, was appropriate for China; but by 1956 there were substantial reservations. The Chinese realized that mechanization could not play the same role in China.

So intimately tied to these large questions was the agricultural mechanization issue that there was a tendency to overlook crucial technical and economic factors. Not until 1957 did agricultural mechanization policy begin to deal effectively with the basic agrotechnical and labor-supply situations in China. Not until 1958 did agricultural mechanization policy relate to economic and cultural transformation within the rural sector. Thus, in a broad sense, the 1950-56 period can be considered a time in which the general elements of development strategy were determined, and a time in which the relevance of the Soviet model to China's situation was considered.

The Soviet Model of Agricultural Mechanization

Before examining China's policy at this time, it will be useful to understand the Soviet model of agricultural mechanization; to a large extent China's policy was first a copy of and then a reaction against it.[1] In the Soviet Union, agricultural policy was very closely tied to agricultural mechanization, which was accomplished through the machine tractor stations (MTS's). One book characterizes their importance this way; "Almost from the beginning, the MTS became a focal point for both economic and political control over the agricultural product and the peasantry."[2]

In the Soviet Union agriculture was collectivized in 1929–30. In implementing collective agriculture, political pressures exerted through cadres, coupled with potential and actual violence, were crucial. However, the economic incentives for collectivization were very weak—in fact they probably were negative—so collective agriculture had a weak foundation. To provide some technological incentive for collective agriculture and to consolidate the collective system, the Soviet leaders choose agricultural mechanization. They reasoned that the economies of scale inherent in field management of tractors would provide a reason for farmers to pool their land and manage it collectively. Even before collectivization, Anastas Mikoyan, speaking at the Second All-Russia Conference of Tractor Workers of Agricultural Cooperation in February 1927, drew attention to the political use of agricultural mechanization:

> This year . . . we want to give all tractors to the cooperatives, the machinery associations, and the collectives. . . . The tractor is not only a good means of mechanized cultivation of the land, but also the best cooperator. It creates the

1. Mao has well characterized the role of the Soviet model: "In the early stages of nationwide liberation, we lacked the experience to administer the economy of the entire country. Therefore, during the First Five-Year Plan, we could only imitate the methods of the Soviet Union, though we always had a feeling of dissatisfaction." Mao goes on to indicate the process by which the Chinese leadership began to question and then depart from the Soviet model beginning in late 1955. Reading Notes on the Soviet Union's *Political Economics, Miscellany,* p. 310.

2. Roy Laird, Darwin Sharp, and Ruth Sturtevant, *The Rise and Fall of the MTS as an Instrument of Soviet Rule* (Lawrence: Governmental Research Center, University of Kansas, 1960), preface. See also Roy Laird, *Collective Farming in Russia* (Lawrence: A University of Kansas Publication, Social Science Studies, 1958), p. 102.

basis for cooperation in the sphere of production. This is the revolutionary significance of the tractor in agriculture. *It is the first and best builder of socialism,* and therefore we attribute great significance to the tractor.[3]

Tractorization at the time of collectivization in the Soviet Union was also important for another reason. One of the ways peasants expressed displeasure with collectivization was by destroying horses and cows. One source estimates that four million horses were destroyed from 1928 to June 1930;[4] another shows that from 1928 to 1933 the number of horses dropped 55 per cent—from 33 million to 15 million head.[5] This rapid, drastic loss of animal power must have created very great problems in plowing. Tractorization was an obvious short-term solution and also a way of reducing the vulnerability of the collective economy to this form of easy sabotage.

The Soviet Union's economic base could not produce tractors instantly at the time of collectivization, but by 1932, only two years after the first major push for collectivization, about one-half of the collectively cultivated land in the Soviet Union was plowed by machine. By 1937 over 90 per cent of the collective land was machine-plowed.[6] (see Figure 3.1).

The degree to which mechanization simplified the establishment and consolidation of collective agriculture is suggested by the fact that on July 1, 1931, in regions in which MTS's had been established, on the average 64 per cent of the peasant farms had been collectivized. In regions outside the MTS zones, only 38 per cent of the farms were in collectives.[7]

3. Cited in Robert Miller, *One Hundred Thousand Tractors—The MTS and the Development of Controls in Soviet Agriculture* (Cambridge: Harvard University Press, 1970), p. 30. Emphasis is Mikoyan's.

4. B. A. Abramov, "The Collectivisation of Agriculture in the R.S.F.S.R.," in V. P. Danilov, ed., *A Survey of the History of the Collectivisation of Agriculture in the Union Republics* (Moscow: State Publishing House of Political Literature, 1963), pp. 103–104; cited in Thomas Bernstein, "Leadership and Mass Mobilisation Campaigns of 1929–30 and 1955–56: A Comparison," *China Quarterly,* no. 31 (July–Sept. 1967), 7.

5. Naum Jasny, *The Socialized Agriculture of the USSR* (Stanford: Stanford University Press, 1949), p. 324.

6. The fact that the Soviet Union collectivized its agriculture before mechanization was widely discussed in the Chinese press. However, the Chinese articles failed to note how rapidly mechanization followed collectivization. Pang Chi-yün, "Coordination between Agricultural Cooperativization and Socialist Industrialization," *Hsüeh-hsi,* no. 12 (Dec. 2), 1955 (ECMM 30, p. 1); Li Yun, "Is Agricultural Cooperativization Possible without Tractors?" *Cheng-chih Hsüeh-hsi,* no. 2 (Feb. 3), 1956 (ECMM 36, p. 18); Tsou Shu-min, "The Role of Advanced Technology in Agricultural Cooperation," *Hsüeh-hsi,* July 1956 (ECMM 50, p. 30).

7. Computed from Miller, *One Hundred Thousand Tractors,* p. 46. The figures for seven regions were simply averaged without adjusting for the differences in size.

Figure 3.1. Kolkhoz plowland: Total area and area serviced by MTS's (million hectares)

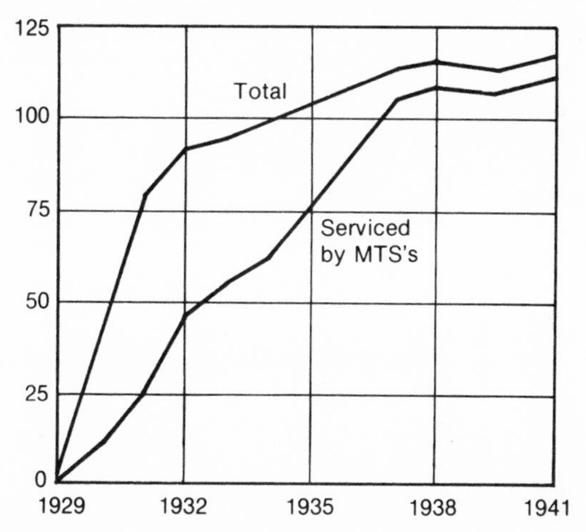

Source: Naum Jasny, *The Socialized Agriculture of the USSR* (Stanford: Stanford University Press, 1949), p. 275. Reprinted with permission.

Agricultural mechanization in the Soviet Union was also used to regulate economic, technical, and political exchanges between the urban and rural sectors. Tractor services were rented out to the kolkhozy, payments being made in grain and pegged at a fairly high rate. The rate was progressive, with a higher percentage for regions with higher yields. In one location in 1947, with a yield of 0.9 to 1.1 tons per hectare, the total payments for machine services—including fall plowing, two cultivations, seeding, and combining—were about 19 to 24 per cent, depending on the precise timing of the service.[8] During the late 1930's these grain pay-

8. Jasny, *Socialized Agriculture,* pp. 290–292. These were percentages of the "biological yield," before harvesting and unavoidable harvesting losses totaling about 23 per cent (Lazar Volin, *A Century of Russian Agriculture* [Cambridge: Harvard University Press, 1970], p. 255). In the 1950's, Malenkov and Khrushchev criticized the system of computing yields on the basis of biological yields, and in 1956 the statistical reporting system ceased using biological yields. Instead of using "barn yields," however (i.e., grain actually in storage), the "bunker weight" seems to have been adopted. This refers to the output collected by harvesting machines and includes some materials which are not suitable for food and some extra moisture. To compensate for these factors, some economists deflate the "bunker yield" by 8 per cent. F. Douglas Whitehouse and Joseph Havelka, "Comparisons of Farm Output in the U.S. and USSR, 1950–71," *Soviet Economic Prospects for the Seventies* (Washington, D.C.: U.S. Congress Joint Economic Committee, 1973), pp. 361–362.

ments constituted a major element in the overall program of procuring grain from the countryside to feed the urban-industrial centers. Payments to MTS's constituted 14 to 19 per cent of total production and over 40 per cent of total procurement, the other means of procurement being through obligatory deliveries, return of loans, and sales to government and on the market.[9] The tremendous importance of MTS payments for state procurement is shown in Table 3.1.

While these charges were high, there is some evidence that they were not exploitative; tractor service was expensive. Jasny crudely estimates that in the late 1930's the MTS system may have had a total income of about 1.4 billion rubles, and total expenses of 1.25 billion rubles. He

Table 3.1. Role of MTS's in grain procurement in the Soviet Union, 1937–40, 1957

Grain procurement	1937–38[a]	1938–39[a]	1939–40[a]	1957[b]
Percentage of crop procured by the state	32.4	38.1	41.5	
Percentage of procurement accounted for by MTS charges	43	42	46	47

Sources:
a. Computed from Jasny, *Socialized Agriculture,* p. 738.
b. Laird, Sharp, and Sturtevant, *The Rise and Fall of the MTS,* p. 58.

concludes that "profit . . . must have been only a minor consideration from the point of view of the state."[10] Thus, even though the MTS may not have been making large profits, it played a major role in procurement of grain and thus was a major element in the regulation of urban and rural economic exchanges.

9. Jasny, *Socialized Agriculture,* p. 738.

10. *Ibid.,* p. 294. Analysts seem confused about the profitability of the MTS system. On the one hand, there is the suggestion that its high profitability was part of a policy to exploit the peasants ("the fiscal importance of MTS as revenue producers was considerable"; Volin, *A Century of Russian Agriculture,* p. 458). On the other hand, it is also suggested that the MTS system was so badly mismanaged that it required constant state subsidies for capital and current expenses. Stalin himself argued that the state subsidies were so large that it was unfeasible for kolkhozy to try to buy and manage machinery themselves. "Reply to Comrades A. V. Sanina and V. G. Venzher," in J. Stalin, *Economic Problems of Socialism* (Moscow: Foreign Languages Press, 1952). This exchange is discussed by Volin, *A Century of Russian Agriculture,* pp. 459–460.

Mao Tse-tung's reaction to Stalin's reply was sharply critical: "Fundamentally, this is incorrect. The basic standpoint is mistrust of the peasants, distrust of the peasants. Agricultural machines were tightly held, and, as a result, the state has the peasants under rigid control. Likewise, the peasants rigidly controlled the state" (*Miscellany,* p. 189).

The MTS system in the Soviet Union was also used for agrotechnical extension. Each station was supposed to have agronomists, experimental personnel, and extension personnel on its staff to test and popularize new cultivation techniques. By 1938, the MTS system employed over 32,000 agronomists; each station could have, on the average, five or six agronomists.[11] This seems to have been the major means of agricultural extension in the Soviet Union at that time; no other extension system existed.

Finally, the MTS system provided a crucial aspect of political control in the countryside. The Russian Revolution had begun in the major cities, and the Communist Party had always been weak in the Russian countryside. In 1929 only 310,000 of the Communist Party's 1.5 million members and candidates (21 per cent), were enrolled in rural party organization;[12] and this was after people began to move from the cities to the countryside. From 1928 to 1930, roughly 250,000 urbanites went to the villages, including about 100,000 Party, government, and economic officials.[13]

After collective agriculture was established, indigenous rural Party leadership was still weak. In mid-1932, only 16 per cent of the kolkhozy had primary Party cells, while 90 per cent of the MTS's had them.[14] As late as the late 1940's, 50 to 85 per cent of the kolkhozy lacked primary Party organizations and were subject to leadership of the local (*raion*) Party committee.[15] To offer political leadership (and control) to their surrounding kolkhozy, political departments were established in each tractor station in 1933, to which some 17,000 Party members were sent. The official history of the Communist Party reports the importance of these political departments:

> In two years (1933 and 1934) the political departments of the machine and tractor stations did a great deal to build up an active body of collective farmers, to eliminate the defects in the work of the collective farms, to consolidate them, and to rid them of kulak enemies and wreckers.
>
> The political departments performed their task with credit: they strengthened the collective farms both in regard to organization and efficiency, trained skilled personnel for them, improved their management and raised the political level of the collective farm members. . . .

11. Jasny, *Socialized Agriculture*, pp. 283–284.
12. Miller, *One Hundred Thousand Tractors*, p. 195.
13. I. Trifonov, *A Survey of the History of Class Struggle in the U.S.S.R. during the Years of NEP (1921–1937)* (in Russian) (Moscow: State Publishing House of Political Literature, 1960), p. 218; cited in Bernstein, "Leadership," p. 29.
14. Miller, *One Hundred Thousand Tractors*, p. 215.
15. *Ibid.*, p. 203.

> [By 1934] since the political departments of the machine and tractor stations had served the purpose for which they had been temporarily created, the Central Committee decided to convert them into ordinary Party bodies by merging them with the district Party Committees in their localities.[16]

Very complex administrative and political relationships developed between the *raion* Party committee, the MTS political department, and the kolkhoz management.[17] While it is difficult to pinpoint the precise locus of leadership, it is clear that the MTS played a major role in consolidating the political control of the Communist Party over the countryside. Being involved in political leadership activities meant being in the vortex of class warfare, directed primarily at the kulaks, and to some extent at the entire peasantry. It is no wonder that in some regions, rifles were part of the standard equipment for tractor station personnel.[18]

These tight economic and political controls over the agricultural sector, in which the MTS played a major role, permitted large extractions from the rural economy of grain, commercial crops, and manpower for urban-industrial development. The rural sector did not benefit. Per capita farm income seems to have declined from 1928 to 1937 by some 10 to 30 per cent.[19] (The extent to which farm income may have been supplemented by off-farm income is not known.)

The Debate over Collective Agriculture

One of the crucial political issues in China in the 1953–56 period was whether collective ownership would emerge as the basic form of rural organization. Land reform had brought important changes. Were they enough? A few years later Mao pointed out the limitations of land reform from a political point of view: "It is by no means strange to distribute land, since MacArthur has distributed land in Japan and Napoleon also distributed land. Land reform cannot elminiate capitalism, and thus will never enable us to reach socialism."[20]

Were more changes, such as collectivization, needed? There was disagreement among Chinese leaders on this question, and it was inevitable

16. *History of the Communist Party of the Soviet Union* (*B*) (New York: International Publishers, 1939), pp. 317, 319.

17. Miller, *One Hundred Thousand Tractors,* chaps. 6–12.

18. *Ibid.*, p. 212.

19. Volin, *A Century of Russian Agriculture,* pp. 257–258. Volin thinks that the only reason the decline may have been as little as 10 per cent was that 1937 was a good harvest year. The larger estimate of decline was made by Naum Jasny, *The Soviet Price System* (Stanford: Stanford University Press, 1951), p. 57.

20. Talk on Problems of Philosophy, Aug. 18, 1964, *Miscellany,* p. 387.

that the Soviet experience with collective agriculture and agricultural mechanization would be brought into the debate. It is remarkable that all parties to the debate used agricultural mechanization as a device to both delay and speed up collectivization. The advocates of collective ownership of land won, but because tractors were not available to consolidate the collective economy, a new plow—the two-wheel, two-blade plow—was used to perform this role. Unfortunately it was not technically suited for many regions in China; it was heavy, cumbersome, and sank in mud. This remarkable attempt to popularize a farm tool primarily for its social consequences will be examined in the following sections of this chapter.

The major arguments in favor of collective ownership were developed by Mao Tse-tung in his speech "On the Question of Agricultural Cooperation" of July 31, 1955.[21] Mao's reasons were not so much ideological as economic and political. From the economic point of view, Mao considered collective agriculture necessary for several interrelated reasons. Through increased scale of operations and better labor organization, collective agriculture would be able to assure increased production. Mao noted that as an empirical observation roughly 80 per cent of experimental cooperatives had increased production.[22] Table 3.2 summarizes some of the surveys upon which Mao relied. The average cooperative member, according to these surveys, had a higher standard of living than the middle peasant. The average family in a cooperative could, at the same time, spend 33 per cent more on production expenses, so that the cooperatives had greater growth potential. A subsample of advanced cooperatives seemed even better off.[23] Because cooperatives had been formed largely by poor peasants without much land, the success was especially impressive. It is not known whether the survey reports were corrected for any state aid that the cooperatives may have received.

The importance of this simple expectation of increased production should not be underestimated. Figure 3.2, based on official Chinese sta-

21. This was released in an English-language pamphlet by the Foreign Languages Press, 1956, which is conveniently reprinted in Robert Bowie and John Fairbank, *Communist China, 1955–1959* (Cambridge: Harvard University Press, 1965), pp. 94–105. A revised translation is included in *Selected Readings from the Works of Mao Tsetung* (Peking: Foreign Languages Press, 1971), pp. 389–420. Citations herein are to this volume, abbreviated *Selected Readings*.

22. *Selected Readings,* p. 399.

23. From *T'ung-chi Kung-tso T'ung-hsün,* no. 17, 1956; cited by Peter Schran, *The Development of Chinese Agriculture, 1950–1959* (Urbana: University of Illinois Press, 1969), p. 124.

Table 3.2. Comparison of income from cooperative and private farming, 1954 (in yuan)

Class of farmer	Family income derived from farming[a]	Total family income[b]	Total income per capita[b]	Production expenses per family	Average income per ha.[a]
Advanced cooperative				234[c]	
All (average) cooperative	466.4	904.20	117.30	107[c]	1.92
Hired laborer and poor peasant	272.6	488.70	116.40	30.3[d]	1.62
Middle peasant	479.7	774.40	154.90	79.8[d]	1.81
Rich peasant	860.6	1,297.00	209.20		1.85
Former landlord	286.0	497.20	118.40		1.49

Sources:

a. Tung Ta-lin, *Agricultural Cooperation in China* (Peking: Foreign Languages Press, 1959), p. 22.

b. *Ibid.*, p. 35. Includes cooperative members' contribution to public funds.

c. From *T'ung-chi Kung-tso T'ung-hsün*, no. 17, 1956; cited by Schran, *Development of Chinese Agriculture*, p. 124.

d. "Simple Data from Surveys on Income and Expenditure of Peasant Families in 1954," *T'ung-chi Kung-tso*, no. 10, 1957; cited by Su Hsing, "The Struggle between Socialist and Capitalist Roads in China after Land Reform," *Ching-chi Yen-chiu*, no. 7 (July 20), 1965 (SCMM 495, p. 13).

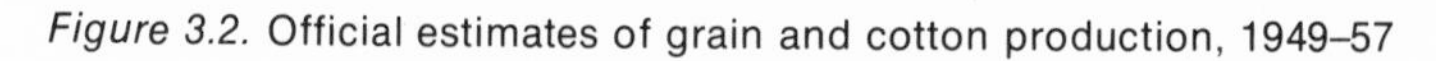
Figure 3.2. Official estimates of grain and cotton production, 1949–57

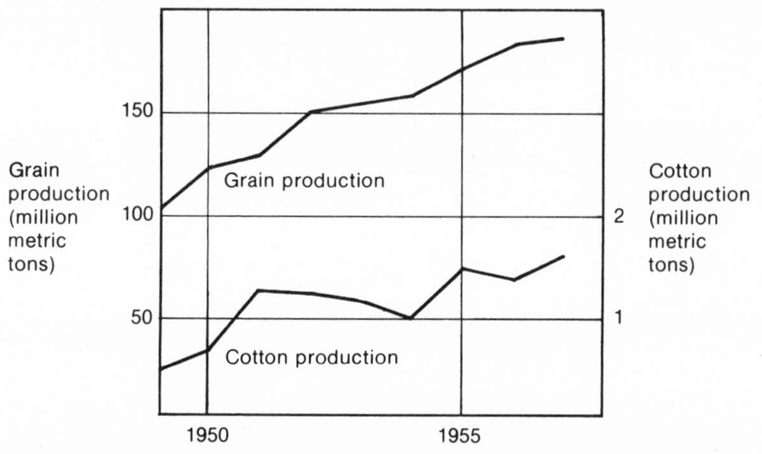

Source: Nai-Ruenn Chen, *Chinese Economic Statistics* (Chicago: Aldine, 1967), p. 338.

tistics, shows that grain production increased after land reform, but in 1953 and 1954 almost leveled off. Cotton production, after a rapid climb until 1951 (due to bringing acreage back to the prewar level), leveled off and actually declined in 1953 and 1954. Whether or not these figures were accurate, they were probably believed by the Chinese leaders, and thus influenced policy.

Su Hsing, a Chinese economist, summarized the situation in 1953 and 1954 in these strong terms: "Agricultural production virtually came to stagnation in the two years under review."[24]

One result of this stagnation was a sharp reduction in industrial growth, after a lapse of one year. Su Hsing noted that "1952 saw a rich harvest in agriculture; and in 1953 industrial production increased by 30.2 per cent, a comparatively great rate. In 1953 and 1954, agricultural production rapidly fell off and the increase in industrial production was considerably slowed down: from 30.2 per cent in 1953 to 16.3 in 1954 and to 5.69 per cent in 1955."[25] (This is, interestingly enough, almost

24. Su Hsing, "The Struggle between Socialist and Capitalist Roads in China after Land Reform," *Ching-chi Yen-chiu,* nos. 7, 8, 9 (July, Aug., Sept.), 1965 (SCMM 495, pp. 1–18; 498, pp. 1–16; 499, pp. 19–33). This statement is at SCMM 495, p. 16.
25. *Ibid.*

the same methodology and conclusions reached by Anthony Tang.)[26]

Kenneth Walker believes that the stagnation in grain production led to a fairly serious grain crisis in late 1954 and early 1955 before the first harvest. Peasants were eating bark off trees and raided government granaries in some places. They also began to migrate to urban centers.[27]

One important implication of the increased production anticipated as a result of collectivization was that more grain and specialized agricultural products would be available to the state for urban and industrial use. Mao

Table 3.3. Extraction of grain from China's countryside, 1954–57 (in million metric tons)

Output and distribution	1954	1955	1956
Total grain output[a]	160	175	183
State grain procurement (tax and purchase)[b]	54	51	49
Grain taken out of rural areas	29	32	25

Sources:
a. Audrey Donnithorne, *China's Economic System* (New York: Praeger, 1967), p. 357.
b. Chou Po-p'ing, "The Policy of Unified Purchase and Sale of Grain Shall Not Be Frustrated," *Liang Shih,* no. 7 (July), 1957 (ECMM 101).

noted: "In the first place, as everyone knows, China's current level of production of marketable grain and industrial raw materials is very low, whereas the state's need for them is growing year by year, and this presents a sharp contradiction."[28] Mao suggested that collectivization would simplify the collection of agricultural taxes and the procurement of commodity grain; during the 1920's the Soviet Union faced the problem of procuring grain and other materials and adopted collectivization as a solution, Mao noted.[29]

However important this perception may have been at the time, in retro-

26. Anthony Tang, "Policy and Performance in Agriculture," in Alexander Eckstein, Walter Galenson, and Ta-chung Liu, eds., *Economic Trends in Communist China* (Chicago: Aldine, 1968), pp. 459–507.

27. Kenneth Walker, "Collectivisation in Retrospect: The Socialist High Tide of Autumn 1955–Spring 1956," *China Quarterly,* no. 26 (April–June 1966), 27. See also Thomas Bernstein, "Cadre and Peasant Behavior under Conditions of Insecurity and Deprivation: The Grain Supply Crisis of the Spring of 1955," in A. Doak Barnett, ed., *Chinese Communist Politics in Action* (Seattle: University of Washington Press, 1969), pp. 365–399.

28. *Selected Readings,* p. 405.

29. *Ibid.*

spect it is clear that establishing cooperatives was not necessary to procure grain and industrial supplies. The government was perfectly capable of extracting grain from the countryside in 1954—indeed, capable of extracting too much. For the purpose of extraction of surpluses, there was no weakness with the system of individual ownership. In fact, it is interesting to note that after the formation of cooperatives, although production rose, the extraction rate declined in both relative and absolute quantities. From 1954 to 1956, state procurement and grain taken out of rural areas went down about 4 million tons, apparently correcting the error made in 1954 (see Table 3.3). Mao explained this error in his article "On the Ten Great Relationships":

> As for the peasants, our relations with them have always been good. But we have made a mistake on the question of grain. In 1954 floods caused a loss of production throughout the country, and yet we purchased 7,000 million more catties [3.5 million tons] of grain. This contrast between reduced production and increased state purchase led to criticism by the peasants. We cannot claim that we never make mistakes. We purchased 7,000 million catties too much because we had no experience and did not have a proper understanding of the problem. This was a mistake. In 1955, having discovered this mistake, we purchased 7,000 million catties less and put the "three stabilizations" into operation. That year there was a bumper crop. Thus an increase in production coincided with a decrease in purchasing, so that the peasants had 20,000 million extra catties [10 million tons] of grain on their hands. All those peasants who had criticized us in the past now no longer criticized us. They all said, "The Communist Party is fine." The whole Party should remember this lesson.[30]

Mao expected that increased agricultural production would increase investment funds available to the government through another mechanism in addition to direct procurement through taxes and sales. He argued that increased productivity would enable peasants to purchase more light industrial consumer goods, such as bicycles, sewing machines, textiles, etc. Prices for these commodities in China were set higher than the cost

30. Mao Tse-tung, "On the Ten Great Relationships," available in Stuart Schram, ed., *Chairman Mao Talks to the People* (New York: Pantheon, 1974), p. 70. Chou En-lai reiterated Mao's views: "However, we also made a mistake in this field [procurement]. In 1954, because of our failure to understand clearly the grain output situation in the country, we purchased a little more grain from the peasants, thus causing the resentment of some of the peasants. In 1955, however, we put into effect the policy of fixed production, fixed purchase, and fixed sale of grain, which pacified the feelings of the peasants and enhanced their enthusiasm for production" ("Report on the Suggestions Regarding the Second Five-Year Plan for the Development of the National Economy," Sept. 16, 1956; available in Chao Kuo-chün, *Agrarian Policies of Mainland China* [Cambridge: Harvard East Asian Research Center, 1957], pp. 23–24).

of production, so that profits could be accumulated. The profits became available to the state for investment, mostly in the industrial sector.

Another serious problem was sociopolitical. After revolution and land reform, rural China was organized on a traditional basis of individual ownership of the means of production—i.e., land and tools. Not surprisingly, by 1953 some of the features of the traditional system began to reappear; the most troublesome of these was inequality.

Table 3.4. Distribution of peasant classes in rural China after land reform and 1954 (in per cent)

Class	At conclusion of land reform	At end of 1954
Poor farm employees	57.1	29.0
Middle peasants	35.8	62.2
Rich peasants	3.6	2.1
Landlords	2.6	2.5
Others	0.9	–
Members of cooperatives	–	4.2

Source: "Simple Data from Surveys on Income and Expenditure of Peasant Families in 1954," cited by Su, "The Struggle," SCMM 498, p. 5.

Table 3.5. Distribution of peasant classes in Hunan, 1950–54 (in per cent)

Class	At time of land reform	1952	1953	1954
Poor peasants	56.73	36.46	28.08	28.22
Middle peasants	30.25	50.45	58.96	58.07
Rich peasants	3.18	3.46	3.63	3.70
Others (landlords, cooperative members)	9.84	9.63	9.33	10.01

Source: Hunan Rural Work Department, "Economic Condition of Poor Peasants in Ts'aot'ang Hsiang, Ch'angsha Hsien," *Jen-min Jih-pao,* Oct. 20, 1955; cited by Su, "The Struggle," SCMM 498, p. 11.

Extensive rural surveys taken by the Chinese in the early 1950's reveal a broad trend toward equality after land reform. Data from a survey of 14,334 peasants in twenty-one provinces in 1954 are reported in Table 3.4. Another survey is reported in Table 3.5. Both of these surveys indicated that the percentage of poor peasants was declining and the percentage of middle peasants increasing. It would seem, therefore, that the overall trend was toward equalization.

The same surveys, however, also uncovered early indicators of trends toward polarization. Some small number of poor farmers were getting poorer, borrowing money, and selling land. There were also scattered reports of small increases in the number of rich peasants. In Hunan the percentage of new rich peasants rose from 3.18 per cent at the time of land reform to 3.70 per cent in 1954. These tendencies were minor and indeed the difference seems so slight that it may not indicate a statistically confirmed trend. Nevertheless the figures worried Communist leaders because they were seen to be the beginnings of a trend toward inequalities, only a couple of years after land reform.

These trends were especially ominous because structural forces were at work to speed up the division of classes into rich and poor. Land reform had not completely equalized peasant holdings. Land expropriations had been limited to the landlords, temples, and clans; most land of rich peasants had not been redistributed. As a result, there was not enough land available for distribution to the small peasants and landless laborers to bring their holdings up to the level of the middle or rich peasants.[31] Peasant families still had different resources and skills; classes still existed.

Random pressures of bad weather and ceremonial needs (marriages, funerals, etc.) quickly forced some of the poor peasants into debt in some localities. In ten *hsiang* (townships) in Hupeh, Hunan, and Kiangsi, a third of all poor peasants were borrowing grain. In Kwangtung, almost half the poor peasants were in debt.[32] At the same time a small group of moneylenders was flourishing. Usurers accounted for twenty households out of 2,486 in seven villages in Shansi.[33] (Perhaps some of these localities had not been self-sufficient, and had depended on remittances or income earned in seasonal migration, both of which were reduced after land reform.)

Another structural factor which reflected and further increased differences was the hiring of labor. A survey of 15,292 households in twenty-two provinces revealed that rich peasants were able to hire poor farm employees to work their farms (Table 3.6). This was considered by some of the leaders to be a continuation of an exploitative relationship that would lead to further impoverishment of the poor farmers.

31. John Wong, *Land Reform in the People's Republic of China* (New York: Praeger, 1973), pp. 158–165.

32. *Selected Reports on Rural Economic Investigations* (Hupeh People's Publishing House), p. 22; cited by Su, "The Struggle," SCMM 498, p. 6.

33. Shih Ching-t'ang et al., *Historical Data on China's Movement for Agricultural Cooperation* (Peking: San Lien Book Shop, 1959), last volume, p. 254; cited by Su, *ibid.*

Table 3.6. Hiring of labor, 1954

Class of family	Avg. man-days of labor hired in by each household	Avg. man-days of labor hired out from each household
Poor farm employees	7.0	25.7
Middle peasants	17.2	18.7
Rich peasants	78.6	11.7
Former landlords	11.8	20.5

Source: "Simple Data from Surveys on Income and Expenditure of Peasant Families in 1954," cited by Su, "The Struggle," SCMM 498, p. 9.

A final structural factor contributing to polarization in the countryside was access to farm tools. Because tools were not integrated into new social structures of mutual assistance and cooperation but were simply redistributed with land during land reform, the traditional unequal distribution of tools soon re-emerged. The rural survey made in 1955 (several years after land reform had been completed and MAT's established) showed that distribution of farm tools were quite unequal (Table 3.7). The rich peasants had about twice as many tools as the poor, and more tools per acre.

Table 3.7. Ownership of land, plows, and waterwheels, 1954

Class of owner	Cultivated land (ha.)	Plows and waterwheels	Tools/ha.
Poor	.83	.48	.58
Middle	1.27	.87	.67
Rich	1.67	1.09	.65
Landlord	.81	.27	.33

Source: "Simple Data from Surveys on Income and Expenditure of Peasant Families in 1954," cited by Su, "The Struggle," SCMM 495, p. 2.

Moreover, there was some evidence that this inequality in tool ownership would continue to grow. The 1954 survey in twenty-three provinces of over 15,000 peasant households showed that the middle peasants had a substantially higher income per household (¥208.4 for middle, ¥ 135.5 for poor farm laborers), and that this higher income permitted much higher expenditures for farm production equipment. Middle-peasant households spent more than twice as much for tools and livestock as did poor farm laborers (see Table 3.2). If this were to continue, the middle peasants would become wealthier, while the poor peasants might stagnate.

Thus, while rural land ownership was much more equalized in 1954 than before land reform, some Chinese leaders feared that there were structural factors in the system of land ownership which would result in greater inequality, and that this had begun already.[34] Mao Tse-tung shared this concern and expressed it clearly in his report "On the Question of Agricultural Cooperation" in the summer of 1955:

What exists in the countryside today is capitalist ownership by the rich peasants and a vast sea of private ownership by the individual peasants. As is clear to everyone, the spontaneous forces of capitalism have been steadily growing in the countryside in recent years, with new rich peasants springing up everywhere and many well-to-do middle peasants striving to become rich peasants. On the other hand, many poor peasants are still living in poverty for lack of sufficient means of production, with some in debt and others selling or renting out their land. If this tendency goes unchecked, the *polarization in the countryside will inevitably be aggravated day by day.*[35]

This situation posed a sharp dilemma for the Communist Party, for in such a situation it would loose its base of support in the countryside. Mao pointed out:

Those peasants who lose their land and those who remain in poverty will complain that we are doing nothing to save them from ruin or to help them overcome their difficulties. Nor will the well-to-do middle peasants who are heading in the capitalist direction be pleased with us, for we shall never be able to satisfy their demands unless we intend to take the capitalist road. Can the worker-peasant alliance continue to stand firm in these circumstances? Obviously not.[36]

While the various reasons for collectivization were spelled out clearly, the arguments against collectivization were never published in a coherent fashion. Most likely, some leaders in China at both the local and national

34. This was not the only problem in rural China at the time. Corruption and laziness of some cadres, commandism, bureaucratism, etc., were dealt with through Party leadership, ideological education, and mass participation. It was not presumed that structural changes in land ownership institutions were needed. Thomas Bernstein, "Problems of Village Leadership after Land Reform," *China Quarterly,* no. 36 (Oct.–Dec. 1968), 1–22; Ezra Vogel, *Canton under Communism* (Cambridge: Harvard University Press, 1969), pp. 104–105. Other questions involved taxes and grain procurement, that is, the economic relations between urban and rural sectors. Likewise, these problems did not call into question the rural institutions. Vivienne Shue, "Taxation, 'Hidden Land,' and the Chinese Peasant," *Peasant Studies Newsletter,* Oct. 1974, pp. 1–12, and "Reorganizing Rural Trade: Unified Purchase and Socialist Transformation," *Modern China* 2:1 (Jan. 1976), 104–134.

35. *Selected Readings,* pp. 411–412. Emphasis added.

36. *Ibid.,* p. 412.

levels thought it desirable that the private ownership system continue. They felt that land reform had eliminated the worst inequalities, and that any change in the ownership principles might be disruptive to production and social order. Many local cadres were opposed to cooperativization, to some extent because they feared peasants would resist change and make leadership much more difficult.[37] At the local level, well-to-do farmers presumably tended to support the private ownership system for reasons of self-interest.

Tung Ta-lin suggests that as late as 1955, there remained considerable opposition to rapid cooperativization even in the Party:

> There were also a handful of right opportunists inside the Communist Party who, echoing the forces of capitalism outside the Party, tried to prevent the broad masses of poor peasants and lower middle peasants from taking the cooperative road. Some of these people, who were communists in name, showed little or no interest in the socialist cause of agricultural cooperation.[38]

Cooperatives and Tractors

While the concerns of various leaders in the early 1950's are understandable, the actual process by which these concerns were integrated and by which authoritative decisions were made remains unclear. How did Mao reach the judgment that collectivization was possible and necessary? Why was most of the Central Committee willing to move only slowly on this question? When Teng Tzu-hui recommended and Liu Shao-ch'i approved the dissolution of 20,000 (some accounts say 200,000) cooperatives in 1955, did this reflect a questioning of the basic principle of agricultural cooperation, or was it simply an effort to control the managerial problems of a rapidly expanding movement? How and why did Mao steer around the Central Committee and take the question to a special meeting which included provincial Party representatives? Why were these provincial-level political leaders more enthusiastic about collectivization than central-level leaders? Why were lower-level (county and *hsiang*) cadres eager to implement the policy?

These questions are not easily answered.[39] Especially perplexing is the

37. In his overall introduction and prefaces to individual articles in *Upsurge of Socialism in China's Countryside* (Peking: Foreign Languages Press, 1957), Mao implied that the local cadres in many regions were very reluctant to go ahead with collective agriculture.

38. Tung Ta-lin, *Agricultural Cooperation in China* (Peking: Foreign Languages Press, 1959), p. 81.

39. A useful chronicle—but not a complete explanation—of this process is Parris Chang, *Power and Policy in China* (University Park: Pennsylvania State University Press, 1975),

question of timing and the role of Chairman Mao. While Mao favored the establishment of cooperatives, his own timetable was remarkably slow. He thought that the spring 1955 plans of the Central Committee to raise the number of lower-level cooperatives with an average of twenty-six households from 650,000 to 1,000,000 in 1956 was a little too cautious. Mao urged that the target be raised to 1,300,000. After a few years of experience, Mao thought that cooperatives would be gradually popularized. By 1958, lower-level, semisocialist cooperatives could be established in most regions. However, roughly fifteen years would be needed for complete socialist transformation.[40] In actuality, China's countryside became collectivized much faster. By the end of 1956, most Chinese peasants were organized into 760,000 cooperatives, most of the advanced type, averaging 155 families. Collectivization of agriculture had been achieved in two years, not the fifteen Mao had predicted.

Somewhere in this debate on collective agriculture, the question of mechanization came up. Just at that time machine tractor stations were beginning to be established, using personnel trained at state farms and tractors imported from the Soviet Union and Eastern Europe. In 1953 eleven MTS's were established on an experimental basis, and by the next year 113 MTS's were serving 40,000 hectares.[41] These stations, patterned on the MTS's of the Soviet Union, rented tractor services to surrounding cooperatives.

From the beginning, the tractor stations established close relationships with agricultural cooperatives to simplify the planning of plowing and to

pp. 9–46. A listing of meetings, which form the skeleton on which a full analysis would be built if data were available, has been compiled by Kenneth Lieberthal, *A Research Guide to Central Party and Government Meetings in China, 1949–1975* (White Plains: International Arts and Sciences Press, 1976).

40. *Selected Readings,* p. 343. Mao repeated the prediction of fifteen years (three five-year plans) in a summing-up speech at the Sixth Plenum of the Seventh Central Committee, Sept. 1955, in *Miscellany,* p. 14. Mao spoke at some length on the dangers of going too fast: "Either too slow or too fast is no good. Both are called opportunism. There is slow opportunism and there is fast opportunism. . . . It is necessary to keep in mind guarding against the 'left.' Guarding against the 'left' is also Marxism. Marxism is not guarding against the right alone. . . . The criterion of quality [in forming cooperatives] is increased production. We don't want dead cattle. [i.e., we don't want peasants to protest as they did in the Soviet Union]. . . . Some people caution us that the Soviet Union previously committed 'leftist' mistakes. We must not forget the experience of the Soviet Union" (Summing-up Speech at Sixth Expanded Plenum of the Seventh Central Committee, Sept. 1955, *Miscellany,* p. 20).

41. "China has 113 Tractor Stations," NCNA Peking, Oct. 29, 1954 (SCMP 919, p. 45).

permit more efficient use of tractors. One of the first large-scale cooperatives—about 200 km. south of Peking, with almost 400 families—was reported to be flourishing, partly because of a nearby tractor station.[42] It was reported that "whole villages and districts have joined cooperative farms in most places within reach of the tractor stations."[43] In Kiangsu, a plan for agricultural development pointed out that "tractor stations must unite closely with consolidation and development of agricultural producer cooperatives, must strengthen their leadership in the cooperativization in the areas."[44]

The observation that tractors would assist agricultural collectivization was, of course, precisely the experience of the Soviet Union, and analogous to the way improved farm tools had helped the mutual-aid teams.

Given the fact that the establishment of agricultural cooperatives was easier with tractors, the question emerged, Did the establishment of collective agriculture *require* tractorization? If so, this would mean virtually an indefinite delay in collectivization of agriculture for technical reasons. At the time of the debate China could not produce any tractors at all; the first tractor factory was designed in July 1953[45] and did not begin mass serial production until 1959. The situation with respect to tractor-drawn agricultural implements was only slightly better. Mass production of such implements began in June 1955.[46] The First Five-Year plan envisaged the creation of an industrial base that could eventually supply industrial inputs to agriculture. But for several years most investment would be in heavy industry. Planned production capacity for tractors in 1959 was only 15,000 units, and only a modest expansion of mechanized state farms and MTS systems was planned.[47]

Rapid mechanization through importation was not a viable alternative. To rely on imports for rapid mechanization would have required huge credits and would have tied up China's foreign trade for decades, assum-

42. "Tractors Work on China's First Cooperative Farm," NCNA Peking, April 11, 1954 (SCMP 786, p. 34).

43. "83 More Tractor Stations Set up in China," NCNA Peking, Jan. 2, 1955 (SCMP 959, p. 13).

44. "Kiangsu Plans to Develop Tractor Stations in Next Two Years," NCNA Nanking, Nov. 16, 1955 (SCMP 1179, p. 18).

45. "China Plans to Build Her First Tractor Plant," China News Service, Peking, June 22, 1954 (SCMP 834, p. 9).

46. "Mass Production of Agricultural Machinery Begins in China," NCNA Peking, June 2, 1955 (SCMP 1062, p. 34).

47. *First Five-Year Plan for Development of the National Economy of the People's Republic of China in 1953–1957* (Peking: Foreign Languages Press, 1956), pp. 78, 126.

ing it were possible to purchase the vast quantities of machinery needed by China, and presuming that markets could be found for China's export commodities. It would also imply extended dependence on other countries in a subtle political sense.

Another major problem with mechanization in 1953–54 was that the technical basis did not exist.[48] Although imported tractors were suitable for reclamation in the Northeast and for cultivating some regions of the North, machinery had not yet been designed for paddy cultivation in the South (Japan was just then developing such machines) or for hilly regions and terraced fields. Implements for specific Chinese crops and intercropping techniques had not yet been designed. It was not known what fuels would be used because at that time China was unable to produce its own petroleum products. For these reasons only about 0.2 per cent of land was plowed by tractors in China in 1952.

Because of the lack of machinery, fuel, productive capacity, designs, etc., Chinese leaders were predicting that the technical transformation of agriculture—at least the mechanization—would take about twenty-five years. In September 1955, Mao predicted that an industrial base would be constructed by 1967 which would be capable of producing 183,000 (standard 15 h.p.) tractors and of providing 18 million tons of crude oil. In another ten years (i.e., by 1977), the technical transformation of agriculture could be achieved, with, among other things, 600,000 tractors.[49]

In retrospect, this was fairly accurate. By 1973, twenty years after the debate, secret documents reported: "The degree of agricultural mechanization . . . is now 27 per cent. Efforts will be made to raise it to 40 per cent by 1975."[50] In 1975, a new plan "to basically fulfill the task of farm mechanization by 1980" was announced.[51]

Despite the lengthy delay in cooperativization that "mechanization first" implied—indeed, I believe that it was precisely because of this lengthy delay—many Chinese apparently argued that cooperative agriculture made little sense until mechanization had advanced. Mao directly

48. This will become apparent later when the wide variety of advances in design and production that were necessary in later years is discussed.

49. Mao Tse-tung, Summing-up Speech at Sixth Expanded Plenum of the Seventh Central Committee, *Miscellany,* p. 16.

50. Political Department, Kunming Military Region, *Reference Materials Concerning Education on Situation,* no. 45, April 6, 1973; available in *Issues and Studies* 10:10 (July 1974), 101.

51. "National Conference on Learning from Tachai in Agriculture," *Peking Review,* no. 38 (Sept. 19), 1975, p. 3.

criticized this view in his report "On the Question of Agricultural Cooperation":

> We are now carrying out a revolution not only in the social system, the change from private to public ownership, but also in technology, the change from handicraft to large-scale modern machine production, and the two revolutions are interconnected. In agriculture, with conditions as they are in our country *co-operation must precede the use of big machinery* (in capitalist countries agriculture develops in a capitalist way). Therefore we must on no account regard industry and agriculture as two separate and isolated things, and on no account must we emphasize the one and play down the other.[52]

Writing in 1958, Tung Ta-lin said, "Some people held that, without tractors, cooperation could not be achieved, or at least it would be slowed down."[53]

During the Cultural Revolution, Red Guards charged that Liu Shao-ch'i, Po I-po, and Teng Tzu-hui favored mechanization first. These charges are difficult to prove from the contemporaneous literature available to us, but it is clear that agricultural mechanization policy was discussed in terms of its relationship to collective agriculture. Liu Shao-ch'i is supposed to have said at different times in 1950: "Without machines, no collective farm cannot [*sic*] be consolidated. . . . Only with the availability of machines shall we be able to run collective farms. . . . Without realizing industrialization, it is basically impossible to realize cooperation in agriculture. . . . Mechanization is a prerequisite to collectivization."[54]

Teng Tzu-hui, director of the Party Rural Works Department, apparently believed at one time that collectivization required mechanization first. Red Guard documents quote him as having asked in 1953: "Agricultural collectivization needs the help of industrialization, for without

52. *Selected Readings,* p. 406. Emphasis added.

53. Tung, *Agricultural Cooperation,* p. 41.

54. Eighth Ministry of Machine Building, Mass Criticism Unit of Revolutionary Great Alliance Committee, Agricultural Machinery Management Bureau, "History of Struggle between the Two Lines (on China's Farm Machinery Front)," *Nung-yeh Chi-hsieh Chi-shu,* no. 9, 1968 (SCMM 633, p. 8). This important source will hereafter be abbreviated "History of Struggle."

This Red Guard accusation rings true. In 1950, in his widely distributed report on the Agrarian Reform Law, Liu Shao-ch'i said: "Only when the conditions mature for the wide use of mechanical farming, for the organization of collective farms and for the Socialist reform of the rural areas can the need for a rich peasant economy cease, and this will take a somewhat lengthy time to achieve." The report was made to the Second Session of the National Committee of the People's Consultative Conference held in Peking, June 1950, and is available in *People's China* 2:2 (July 16, 1950), 5–9, 28–31. The quotation is at pp. 28–29.

tractors, how can we bring about collectivization?''[55] By September 1954, however, Teng seems to have changed his emphasis. In a report on the socialist transformation of agriculture to the National People's Congress, Teng observed that

> mechanization of our agriculture should be based on collectivization and along socialist lines and not along capitalist lines.
>
> However, the fact that our country is not able to make tractors now and production of gasoline is low while the socialist consciousness of the peasants can not be heightened overnight, dictates to us the necessity of effecting socialist transformation of agriculture in two stages: the first stage is socialist revolution to achieve cooperativization; the second is technical revolution to attain large-scale mechanization.[56]

Teng went on to point out that both the political and technical transformation of agriculture would be done in two steps. Socialist transformation would involve the development of MAT's to cooperatives and then to collectives. The mechanization would begin with semimechanization, i.e., horse-drawn equipment.

Po I-po was also charged by Red Guards with opposing early cooperativization. In an article published in *People's Daily,* November 17, 1955, he seemed to support gradual collectivization: ''Under the conditions of our country, without cooperation it is impossible to effect technical reform of agriculture. For modern agrotechnology is based on mass production; without enlarging the production scope of agriculture through introduction of cooperation, it is impossible to adopt modern agrotechnology.''[57] While this argument does not mean ''machines before cooperation,'' it suggests there need be no rush for cooperation until modern inputs are available. Po (accurately) estimated that agrotechnical reform in China would require twenty to twenty-five years. Cooperatives should come before tractors—but just a bit before. This seems to be in cautious agreement with Mao's enthusiastic call for cooperatives about three months earlier. At this time Po was sensitive to the problem of what to do with labor displaced by machines; he advocated expansion of industry, agriculture, transportation, communications, education, and culture to absorb the manpower.

T'ao Chu, first secretary of Kwangtung Party, made a report to the

55. ''History of Struggle,'' SCMM 633, p. 8.

56. Teng Tzu-hui, ''Report on Socialist Transformation of Agriculture'' (to the National People's Congress), Sept. 23, 1954, NCNA Peking (SCMP 898, p. 10).

57. Po I-po, ''Agricultural Cooperation Should Be Closely Linked up with Technical Reform of Agriculture,'' *Jen-min Jih-pao,* Nov. 17, 1955 (SCMP 1179, p. 36).

First Conference of Representatives of the Chinese Communist Party (CCP), on October, 3, 1953, in which he tactfully skirted the issue of what social changes could be achieved before mechanization:

> Today we do not have large quantities of machinery equipment which can be supplied to agriculture, and *we cannot simply wait for such machinery equipment.* But there are many tasks which we can now perform. As an instance, the agricultural productive potential in China is still very large, and up to now we have not put to the fullest employment our existing manpower and our present techniques. There is still a period of time during which we may exert to the fullest extent our present potentialities, and by that time we shall be in possession of machinery equipment and agriculture may then be basically reformed. . . . There must be extensive development and gradual elevation of temporary seasonal mutual aid teams, and the fostering and steady extension through keypoints of permanent mutual aid teams. There must be *keypoint experiments on agricultural producers' cooperatives,* and the planned construction, during the period of the five year plan, of the few state farms which can really be looked upon as examples to the people.[58]

T'ao seemed to say that some social changes could be made before mechanization. But he did not suggest that cooperatives could be established on a universal basis.

The precise content of the debates in 1954 and 1955 over the suitability of agricultural cooperatives is not known, but it seems clear that the debate was argued (to some extent at least) not in terms of the desirability of cooperation but rather in terms of the relative timing of cooperation and mechanization. Apparently no one—at least no politically sophisticated person—dared to oppose agricultural cooperation as a matter of principle. The idea of agricultural cooperation was so intimately associated with socialist ideology and with the Soviet model of development that any direct opposition would have been considered "counter-revolutionary." The "mechanization first" argument, however, could provide an elegant way to oppose agricultural cooperation. Based on Marxist historiography that emphasizes the primacy of the means of production, this convenient argument would require a virtually indefinite delay for cooperativization.

As top leaders recognized the enormity of the technical problems in ag-

58. T'ao Chu, "The Tasks of the Party Organization in South China during the State's Five-Year Construction Plan" (a report to the First Conference of Representatives of the CCP, Oct. 3, 1953), *Nan-fang Jih-pao,* Oct. 31, 1953 (SCMP 703 supplement, p. iv). Emphasis added.

Table 3.8. Rural Party organization on the eve of collectivization, China and the Soviet Union

Party membership	USSR 1929	China 1955	China 1956
Percentage of Party members in rural areas	21%[a]	43%[b]	70%[b]
Rural Party members (thousands)	310[a]	4,000[b]	7,517[b]
Total rural population (millions)	114[c]	532[d]	539[d]
Rural Party members per 1,000 rural inhabitants	2.7	7.5	13.8
Percentage of villages with Party organization	33–39% of 72,163 villages[e]	77% of 220,000 *hsiang*[b]	90% of 210,000 *hsiang*[e]

Sources:

a. Miller, *One Hundred Thousand Tractors,* p. 195.

b. Franz Schurmann, *Ideology and Organization in Communist China* (Berkeley: University of California Press, 1968), pp. 129–32.

c. Frank Lorimer, *The Population of the Soviet Union* (Geneva: League of Nations, 1946), p. 110; cited in Jasny, *Socialized Agriculture,* p. 709.

d. Schran, *Development of Chinese Agriculture,* p. 47.

e. Bernstein, "Leadership," p. 11.

ricultural mechanization, as they realized that "mechanization first" meant the indefinite postponement of socialist agriculture, and as they observed the problems in the rural sector in 1953–55, they seem to have adopted the view that agricultural collectivization was necessary.

The smooth, rapid collectivization of China's countryside—without the incentives inherent in tractorization—was possible because of several factors. First, the Chinese Communist Party had a firm rural organization by 1955. Table 3.8 shows how much stronger the CCP was in rural areas than was the Communist Party of the Soviet Union on the eves of their respective collectivization drives. In China, the ratio of Party members per rural inhabitants was roughly three times that in the Soviet Union; in China the rural Party organization had been constructed in almost 80 per cent of the rural townships (*hsiang*) while in the Soviet Union the figure was about 40 per cent.

Second, the Party in China exhibited much more effective leadership. Its cadres were not only more numerous but more familiar with rural conditions and more reliable. The Chinese were much more careful to develop local, indigenous leadership and to mobilize the lower classes of peasants to neutralize the middle and rich peasants.[59] At the same time the Chinese leaders had, over the previous five years, through careful, deliberate manipulation of agricultural taxes, credit, procurement, and marketing policy, created a system of economic incentives in which the middle and rich peasants found participation in collective activities financially desirable.[60] By 1955, about half of China's rural population had been in mutual-aid teams, and about 14 per cent had been in lower-stage cooperatives.[61]

Another factor simplifying the establishment of collective agriculture in China was the long tradition of collective activities. At the village level, sharing of tools and exchange of labor had long been a tradition.[62] At the multi-village level, in many regions there was a long experience of collective management of water.[63] The market town, comprising perhaps twenty villages and a population of seven or eight thousand people, was a well established social community with substantial social intercourse (through kinship and economic networks) and with mechanisms for sharing tasks and benefits.[64] These various traditional mechanisms meant that the Chinese peasant had experience working with other people. The notions of reciprocity had been well established. There was a social, psychological, and economic foundation on which agricultural cooperatives could be built. (Perhaps the Soviet Union had a similar foundation for collective agriculture in the *mir,* but it was not used in the same way.)

The decision to establish collective agriculture in China long before tractorization would be possible left a crucial constraint on the Chinese political system. Since there was no technological basis for economies of scale in agriculture, peasants might tend to leave collective agriculture and "go it alone." The political system has frequently had to exert its leadership and has had to assure that rural standards of living would go up to keep peasants in collective agriculture. In the Soviet Union, in contrast, rapid mechanization of field crops made withdrawal from the kol-

59. Bernstein, "Leadership," pp. 17–47.

60. Vivienne Shue, "Transforming China's Peasant Villages: Rural Political and Economic Organization, 1949–56" (Ph.D. diss., Harvard University, 1975).

61. Bernstein, "Leadership," p. 17.

62. John Wong, "Peasant Economic Behavior: The Case of Traditional Agricultural Cooperation in China," *Developing Economies* 9 (1971), 332–349.

63. Ramon Myers, "Economic Organization and Cooperation in Modern China: Irrigation Management in Xing-Tai County, Hobei Province," in *The Polity and Economy of China* (Tokyo: Toyo Keizai Shinposha, 1975), pp. 189–212.

64. G. William Skinner, "Marketing and Social Structure in Rural China," *Journal of Asian studies* 24 (1964), esp. 35–43.

khoz virtually impossible. The Soviet system has been able to maintain high rates of extraction and keep rural incomes depressed without fearing that the collectivization of field crops would dissolve. Even if peasants put much energy into private plots in the Soviet Union, the collective production of crucial grains is assured. The Chinese leadership has been well aware of this constraint on its policy toward the rural sector, especially since the serious depression of 1959–61.

New Plows and Agricultural Cooperatives

Just as the people who opposed agricultural collectivization based arguments on the lack of farm machinery, so too those who favored collectivization fully realized the validity of the argument that new tools would simplify the evolution of enlarged farming. Since tractors could not be supplied quickly enough for cooperative agriculture, a new type of plow—with two wheels and two blades—was proposed for rapid extension to give cooperative farming a suitable technical basis. The socializing effect of the two-wheel, two-blade plow had been noticed in December 1953, when just a few thousand had been produced. NCNA reported: "Many individual peasants have decided to form mutual-aid teams so that they may afford to buy new-type plows. . . . Mutual-aid teams are pooling their land into cooperative farms so as to make fuller and more efficient use of new-type equipment, particularly entire sets of horse-drawn machines."[65]

The new plows thus offered a solution to the nagging problem of assuring that the cooperatives would have a sound economic basis. To assure high and constant rates of extraction of grain and industrial crops within the framework of the principle of voluntary membership and individual benefit, an increase in production and productivity was required. Although there was some survey data indicating that group labor would be more effective, many Chinese leaders thought that new cultivation techniques or agricultural inputs would be necessary to raise production dramatically.[66] The two-blade, two-wheel plow was the answer, they

65. "New Type Farm Equipment Helps Raise Crop Yields in China," NCNA Peking, Dec. 19, 1953 (SCMP 712, p. 33). See also "Chinese Peasants Use New Type Animal-Drawn Farm Tools," NCNA Peking, Aug. 1, 1954 (SCMP 862, p. 27).

66. A conference to review their use in July 1954 reported that the new implements (including plows, cultivators, harrows, etc.) raised yield by 15 per cent in the Northeast. "Chinese Peasants Use New Type Animal-Drawn Farm Tools," NCNA Peking, Aug. 1, 1954 (SCMP 862, p. 27).

thought. One Chinese analyst put this argument forcefully in an article published in July 1956:

> To rapidly promote agricultural cooperativization, the peasants' needs of modern farm tools and other items must be satisfied. Otherwise the superiority of agricultural producer cooperatives will not be noticed so easily.
>
> If an agricultural producer cooperative [APC] does not have new techniques, but merely relies on the improvement of labor organization, it will be difficult to increase rapidly year after year the production of the unit and the income of its members. If the original APC's fail to display their marked superiority, the peasants will have doubts over the cooperatives, and it will be more difficult to promote rapidly the development of agricultural cooperativization.
>
> Since *new techniques* play a definite role in th the consolidation and development of agricultural producer cooperatives, after the arrival of the high tide of socialist revolution in the countryside last year the peasants have more urgently demanded modern farm tools and agricultural machinery. According to the national plans for the current year, the supply of *double-wheel double-blade* and *double-wheel single-blade* plows this year will be more than five times that of last year. These figures reflect that the further development of agricultural cooperativization still relies on the support of industry.[67]

In addition to providing a technical reason for fully pooling land, the new plow was claimed to increase yield by 10–15 per cent.[68] Thus it would offer an opportunity for the cooperative to raise the income of its members.

In late 1955 and early 1956 plans were made to expand production rapidly so that by the end of the year the plow would be the major tilling instrument in China. Initial plans for 1956 production were developed at a conference in November 1955, which decided to raise the number of wheeled plows planned for distribution in 1956 to roughly 800,000, double the number distributed in 1955.[69] On December 10, 1955, the target was again raised: 2,360,000 two-wheeled plows were to be produced and 1,800,000 of them distributed in 1956. Plans were also developed for training people to use and repair the plows, and repair centers were

67. Tseng Wen-ching, "How Will Successful Socialist Industrialization Promote the Development of Agricultural Cooperativization," *Cheng-chih Hsüeh-hsi,* no. 7 (July), 1956 (ECMM 51, p. 18). Emphasis added.

68. "Peasants Buying More New Type Wheel Plows," NCNA Peking, Nov. 1, 1955 (SCMP 1162, p. 22).

69. "Machine Building and Chemical Industrial Departments Actively Prepare Extended Production of Farming Implements and Fertilizer Next Year," NCNA Peking, Nov. 12, 1955 (SCMP 1177, p. 6).

Table 3.9. Number of two-wheel, two-blade plows required for nationwide adoption

Type of area	Size of area (million ha.)	Days of plowing season	Ha./plow	Plows needed (millions)
Principal computations				
Dry land	63	20	13	4.0
Wet land	27	20	10	2.7
Total				6.7
Alternative computations				
Dry land	50	20	17	3.0
Wet land	23	20	11	2.2
Total				5.2

Source: Planned Economy editor, "Why Does the Demand for Double-Wheel and Double-Blade Plows Drop and Why Is Their Production Suspended?" *Chi-hua Ching-chi,* no. 9 (Sept. 23), 1956 (ECMM 56, p. 24).

projected.[70] In January 1956 the production goal was raised again, to 4,000,000, of which 800,000 were to be distributed in the first quarter.[71] Four million plows were enough to make the plow the major tilling instrument in 65 to 85 per cent of China, as Chinese planners projected total national need at 5.2 to 6.7 million (Table 3.9). High priority was attached to this plan. Municipal and provincial Party and government leaders were asked to "pay much attention," and to "personally undertake to organize production and to supervise examination." Three work teams—with representatives of the First and Third Ministries of Machine Building, and the Ministry of Agriculture, as well as engineers and trade union leaders—were sent to nine provinces to supervise production of the plows.[72]

Although provincial and municipal authorities were hoping that the new plows would provide the technical underpinning for the new cooper-

70. "More Modern Farm Tools to be Produced Next Year by Factories throughout the Country," NCNA Peking, Dec. 10, 1955 (SCMP 1192, p. 8); "Third National Farm Tool Conference Decides to Extend More New Type Farm Tools Next Year," NCNA Peking, Dec. 15, 1955 (SCMP 1194, p. 12).

71. "Great Increase in Supply of Farm Tools," NCNA Peking, Jan. 19, 1956 (SCMP 1214, p. 27).

72. "Farm Tool Makers to Produce 800,000 Units of Double-Bladed and Single-Bladed Wheel Plows during First Quarter," NCNA Peking, Feb. 1, 1956 (SCMP 1234, p. 8).

atives, it is clear that these decisions were based on hopes and not on field experimentation. As soon as the plow was distributed in southern China, serious technical deficiencies were noted. It sank into the mud of paddy fields and was too heavy and cumbersome to be used on the small plots of paddy or terraced fields. It required more draft power than was available in many places. Generally two beasts were required to pull it, and the Chinese water buffalo refused to work in teams.[73] It was too wide for the paths between fields.[74] Moreover, parts and repair service were not yet available in areas where the plow had not been distributed previously. Finally, in August 1956, production of the plow was temporarily suspended.[75]

Chinese Communist leaders knew how to avoid this fiasco. In most movements, including land reform and the establishment of cooperatives, test sites were established, experience was gained in a particular locality, and modifications were made in the programs to fit local conditions and requirements. Why was this not done for plows? Apparently agricultural mechanization in China before 1957 was considered simply in terms of how it might reinforce collective agriculture. Little attention was given to the technical side of changes in agricultural technology. Chinese leaders did not seriously question what kinds of machines would be produced, how they might differ to meet differing local needs, how machinery would relate to cropping patterns, industrial organization, or even labor supply. Even though they were sensitive to the nuances of social change, they vastly underestimated the problems of technical change.

The result was a policy that was a blunder. The two-bladed, two-wheel plow wasted steel and other industrial resources, and undoubtedly embarassed the political authorities who overemphasized it. The plow did, however, have the potential of becoming a useful negative example. In a general sense it demonstrated that agro-technical change is complex, and requires careful planning, local experimentation, regional diversity, and appropriate training of personnel. If changing plows was difficult for the Chinese political system, how much more complex would be coordinated changes in seeds, fertilizers, and cultivation methods? In short, the expe-

73. Kenneth Walker, "Organization of Agricultural Production," in Eckstein et al., eds., *Economic Trends in Communist China,* p. 420.

74. "Hunan Examines Extension of Double-Bladed Wheel Plow," NCNA Changsha, June 1, 1956 (SCMP 1312, p. 17).

75. *Planned Economy* editor, "Why Does the Demand for Double-Wheel and Double-Blade Plows Drop and Why Is Their Production Suspended?" *Chi-hua Ching-chi,* no. 9 (Sept. 23), 1956 (ECMM 56, p. 24).

rience revealed that the Chinese political system was much more sensitive to the social implications of technical changes than to the technical aspects of those changes.

By 1957 other factors also permitted mechanization policy to proceed on a new, more realistic basis. First, agricultural cooperatives were established; thus mechanization could become separated from the debate over agricultural collectivization. Second, an industrial base was being built, so it became more realistic to talk about what kind of agricultural machinery to produce. Third, through the state farms and the machine tractor stations a certain amount of experience had been gained concerning the process and implications of agricultural mechanization.

While critical obstacles to the formulation of an effective mechanization policy were thus removed, there was not yet widespread agreement among the Chinese leadership on two crucial aspects of mechanization policy, namely the rate of mechanization and the institutional patterns. On the first point, many in China still felt that in a densely populated country with an intensive agricultural system, mechanization was not too important and might result in unemployment. On the institutional issues, there was at first agreement that the Soviet style MTS system would be appropriate. Soon, however, debate emerged on this question; experiments were tried with alternative institutions, and argument intensified. The next chapter will examine the early course of the debate on institutions, and the following chapter will look at the overall question of fitting mechanization into a strategy of agricultural development.

4. Mechanization and the Distribution of State Power

When the Chinese leaders decided that collectivization could proceed before mechanization, they did not disparage the importance of agricultural mechanization, as implicit in the Soviet model. As in the Soviet Union, machine tractor stations were expanded to mechanize agriculture and to consolidate the collective economy.

In a few years, however, the Chinese leaders decided that the MTS was inappropriate. It represented a centralized, bureaucratic pattern of organization which had tendencies toward elitism and inefficiencies. By 1957 experiments were underway to break apart the bureaucracy and decentralize ownership of machinery to newly established communes. Instead of being a mechanism of state political and economic control over the rural sector, agricultural mechanization would eventually become a new cultural and economic force within the rural society.

When machine tractor stations were established in February 1953, they were presumed to "pave the way for mechanization of agriculture in the future."[1] Their number and resources grew quickly. By 1956, MTS's possessed over 50 per cent of the nation's tractors (Table 4.1). After an experiment with an alternative form of ownership—commune ownership in 1958–59—the MTS (renamed agricultural machinery station, AMS) became even more important. In 1965, the AMS's owned over 60 per cent of the nation's tractors.[2] At that time, the AMS system owned almost 80,000 standard tractors, about eight times what it owned in 1956. Because of problems with the AMS system, it was basically dismantled after the Cultural Revolution.

1. "Peking Tractor and Farming Machinery Station Begins Spring Work," NCNA Peking, March 29, 1953 (SCMP 543, p. 19).

2. At some risk of confusion, this chapter deals with organization of both the MTS's in the 1950's and the AMS's in the 1960's because the systems were substantially similar. Differences are examined in Chapter 8.

Table 4.1. Machine tractor/agricultural machinery stations, 1950–65

Year	Number of stations	Standard tractors owned by stations	% of total tractors [a]	Tractors/ station	Land serviced	Ha./ tractor [a]
1950	1	30	2	30		
1951	1	30	2	30		
1952	1	30	1	30		
1953	11	68	3	6.2		
1954	89	778	15	8.9		
1955	138	2,363	29	17.1	226,000 ha.[b]	102.7
	(139) [b,c]	(2,203) [b]				
1956	326	9,862	51	30.3	1,350,000 ha.[c]	150.0
	(325) [c]	(9,600) [c]				
1957	383	12,036	49	31.5	1,746,000 ha.[d,e]	145.3
1958		(10,995) [f]	(22) [f]			
1959	553	(17,300) [f]	(29) [f]	31.3 [g]		
1963	1,482	68,040	59	46		
1964	1,488	71,500	58	48		
1965	2,263	79,300	61	35		

Sources: Except where otherwise indicated, source material for this table comes from Kang Chao, *Agricultural Production in Communist China, 1949–1965* (Madison: University of Wisconsin Press, 1970), pp. 109, 111. Exceptions are listed below:

a. Computed.

b. "Good Results Achieved in the Construction of Agricultural Machine and Tractor Stations in the Country," NCNA Peking, Jan. 10, 1956 (SCMP 1217, p. 14).

c. "325 Tractor Stations in China," NCNA Peking, Feb. 26, 1957 (SCMP 1480, p. 10).

d. "More Tractors for Agricultural Cooperatives," NCNA Peking, Jan. 18, 1958 (SCMP 1697, p. 15).

e. "Tractor Station Conference," NCNA Peking, Feb. 16, 1958 (SCMP 1719, p. 4).

f. These figures are Chao's, but it should be noted that they are estimates, and subject to error.

g. Assumed by Chao.

While the geographic distribution of the tractor stations is not known precisely, many of them were in the Northeast and North China Plain. Available data are presented in Table 4.2.

Organization of the MTS/AMS System

The Chinese had several sources for the idea of machine tractor stations. Some of the initial practices of the MTS system were probably derived from the United Nations Relief and Rehabilitation Administration

Table 4.2. Location of machine tractor stations, 1956–57 (incomplete)

Region and province	1956	1957
Northeast		
Heilungkiang	add 10[a]	28[b]
Kirin		12[c]
Liaoning	19[d]	23[c]
North-central		
Hopei	45[e]	(45)
Shansi		
Inner Mongolia	4[f]	(4)
Northwest		
Shensi	1[g]	
Kansu		(1)
Ninghsia		
Tsinghai		
Sinkiang	2, *P*5[h]	(7)
East-central		
Shantung	38[i]	(38)
Kiangsu	22[j]	*P*43[j]
Anhwei	13[k]	(13)
Chekiang		
Fukien		
Shanghai	*P*3[l]	(3)
Central-south		
Hupeh		
Hunan		
Honan		40[m]
Kwangsi	6[n]	9, *P*12[n,o]
Kwangtung		
Kiangsi	8[p]	(8)
Southwest		
Szechwan		
Kweichow	1[q]	(1)
Yunnan	1[r]	(1)
Sikang		
Number located	160	279
National total	326	383

Note: Figures in parentheses are estimates based on the assumption that the number of stations reported in existence or planned in 1956 existed during 1957. This may not be a valid assumption, because additional stations may have been constructed. *P* indicates that the specified number of stations were planned.

Sources:

a. "Tractor Stations Set up for Land Reclamation in Heilungkiang," NCNA Harbin, March 15, 1956 (SCMP 1251,

p. 13); "add" indicates that the specified number of stations were added to whatever existed before.

b. "Heilungkiang Plans for Farming Mechanization," NCNA Harbin, Feb. 15, 1958 (SCMP 1717, p. 7).

c. "Agricultural Mechanization Bears Fruit in Northeastern 3 Provinces," NCNA Harbin, March 18, 1958 (SCMP 1740, p. 8).

d. "More Tractor Stations for Northeast Province," NCNA Shenyang, Jan. 7, 1956 (SCMP 1205, p. 21).

e. "Hopei Builds More Tractor Stations, NCNA Peking, Jan. 8, 1956 (SCMP 1205, p. 20).

f. "New Tractor Stations Set up in Inner Mongolia," NCNA Huhehot, July 20, 1956 (SCMP 1336, p. 14).

g. "Yenan Tractor Station Starts Autumn Plowing," NCNA Yenan, Oct. 17, 1956 (SCMP 1394, p. 7).

h. "More Tractors for Sinkiang," NCNA Urumchi, March 1, 1956 (SCMP 1241, p. 9).

i. "More Tractor Stations in Shantung," NCNA Tsinan, March 2, 1956 (SCMP 1242, p. 13).

j. "Kiangsu Plans to Develop Tractor Stations in Next Two Years," NCNA Nanking, Nov. 16, 1955 (SCMP 1179, p. 18).

k. "School for Mechanized Farming in Anhwei," NCNA Hofei, Oct. 8, 1956 (SCMP 1388, p. 19).

l. "Farming in Shanghai Area to be Mechanized," NCNA Shanghai, Jan. 5, 1956 (SCMP 1204, p. 27).

m. "An Epitome of Agricultural Development in Honan in the Past Five Years," *Honan Jih-pao,* Chengchow, April 6, 1958 (SCMP 1771, p. 46).

n. "Kwangsi Province to Set up 300 Tractor Stations within 3 Five-Year Plans," *Ta Kung Pao,* Nanning, Nov. 11, 1956 (SCMP 1420, p. 24).

o. "State-owned Mechanized Farms in Western Kwangsi Chuang Nationality Area," NCNA Nanning, Dec. 24, 1955 (SCMP 1207, p. 34); "Six New Tractor Stations for Kwangsi," NCNA Nanning, March 25, 1956 (SCMP 1257, p. 18).

p. "Tractor Stations in Kiangsi Make Profit as Result of Industry and Thrift," NCNA Nanchang, Jan. 24, 1958 (SCMP 1714, p. 23).

q. "Kweichow Establishes First Tractor Station," NCNA Peking, March 29, 1956 (SCMP 1261, p. 31).

r. "Yunnan Tractor Station Starts Autumn Plowing," NCNA Kunming, Oct. 20, 1956 (SCMP 1395, p. 16).

(UNRRA). At least one of the UNRRA technical experts in agricultural machinery, William Hinton, remained in China after 1949 and trained some of the tractor hands.[3]

3. William Hinton, *Iron Oxen* (New York: Vintage 1971).

The most important influence on the establishment of MTS's was the example of the Soviet Union, where the MTS system was the dominant form of tractor ownership.[4] As early as 1950, the Chinese were studying the Soviet system and reporting its advantages.[5] Not only were Soviet tractors important in the initial stocking of the Chinese MTS's;[6] Soviet (and Hungarian) specialists taught the first group of cadre and staff how to operate the equipment. In addition, many Chinese drivers, mechanics, and technicians were trained at the Sino-Soviet State Friendship Farm. The Soviet Union also supplied a great deal of the institutional framework, such as management procedures, standards to assure proper maintenance of machines, and guidelines for the evaluation and reward of personnel.

It need not be assumed, of course, that the Chinese were simply imitating tractor stations that had been invented by the Soviet Union or UNRRA. They may have independently perceived the technical advantages of tractor stations, namely that concentration of tractors and operators under strict control of the state permits careful supervision over utilization and maintenance of machinery and assists in training of personnel. These factors are particularly important in introducing machinery into a country which lacks both people who are familiar with machinery and an industrial infrastructure to service machinery. (It was for these reasons that the UNRRA projects kept machinery concentrated in units of roughly 20 tractors.)[7]

During the 1950's the structure of the tractor station system was simple. Each station had, on the average, about 30 standard 15-h.p. tractors. In the 1960's, however, as the system grew, a more sophisticated three-level management structure was suggested.[8] (Whether or not it was actually implemented is not known.) At the county level, there was to be a

4. Excellent surveys of the organization of the MTS system in the Soviet Union are Robert F. Miller, *One Hundred Thousand Tractors* (Cambridge: Harvard University Press, 1970); and Roy Laird, Darwin Sharp, and Ruth Sturtevant, *The Rise and Fall of the MTS as an Instrument of Soviet Rule* (Lawrence: Governmental Research Center, University of Kansas, 1960).

5. Li Chi-hsin, *Nung Chi Chan* [Agricultural Machinery Stations] (Shanghai: Commercial Press, 1950).

6. "Peking Tractor and Farming Machinery Station Begins Spring Work," NCNA Peking, March 29, 1953 (SCMP 543, p. 19).

7. Interview with Irving Barnett, Political Science Department, State University of New York at New Paltz, who worked with the UNRRA agricultural program.

8. Chang Ch'ing-t'ai, "On the Question of Strengthening the Operation and Management of Tractor Stations," *Chung-kuo Nung-pao*, no. 11 (Nov. 10), 1963 (SCMM 400, p. 31).

general tractor station, which would supervise mechanization activities. Under it would be district (*chü*) or large-commune tractor stations. (There are roughly five to ten districts in a county.) Each station would serve perhaps 10,000 to 20,000 ha. within a radius of 10 km. It would own perhaps 30 to 50 large tractors (the equivalent in power of 60–100 standard 15-h.p. tractors). The tractor station would be the basic accounting unit.

The actual tasks were to be carried out by mechanized farming teams. A station would have four to ten teams. Each team would have four to ten tractors (10–15 standard tractors), and service 2,000 ha. within a 3-km. radius. Generally one team would be set up for a commune, but large communes might have several teams.

Areas that were not advanced in mechanization would have a two-level system of management, eliminating the middle level.

Leadership

The machine tractor stations—as other units in the Chinese Communist economic system—were subject to dual leadership: government and Party. From the point of view of government administration, during the 1950's the MTS's were under a special bureaucracy called the Ministry of State Farms and Reclamation (Nung K'en Pu) at the national level, which had an office at the provincial level (Nung Kuan T'ing). Presumably each county also had such an office, which directly supervised the MTS's. From 1959 to 1965 the Ministry of Agricultural Machinery provided leadership, and after January 1965, the Eighth Ministry of Machine Building supervised the newly renamed agricultural machine stations. The Ministry of Agriculture, which was responsible for overall production and for achievement of collectivization, did not have primary responsibility for supervising agricultural mechanization. One major function and power of the administrative system was to control the careers—including job rank, salary, and location of work—of the stations' employees. The government administrative system seemed primarily concerned with the internal efficiency of the stations and was, apparently, only marginally and indirectly concerned with how they related to surrounding areas.[9]

To judge the internal efficiency and general appropriateness for a particular location of a tractor station the administrative bureaucracy used financial accounting of profit and loss of each station. The accounting

9. This paragraph is based on interviews with people who worked in China.

procedures are not known for certain, but one economist suggested a detailed breakdown of costs in 1964:[10]

Variable costs, according to acreage

1. Fuel
2. Maintenance and repairs

Fixed costs

3. Wages and extra wages for workers
4. Wages and extra wages for administrators
5. Reserves for major overhauls
6. Depreciation of fixed assets
7. Operation and management expenses

It is not clear whether depreciation was included in computations before 1957. It was supposed to be, according to Soviet texts that were used for training; but an instruction issued at the end of 1957 stated: "In the next one or two years, all stations are required to calculate depreciation cost on fixed assets."[11] This seems to suggest that depreciation was not included before 1957. In the 1960s the accounting system was still not counting depreciation properly. Stations sometimes declared machinery "obsolete" prematurely so they could write it off the station's inventory and not be billed for depreciation. However, the machinery might still be utilized by the station.[12] For example, one station in Heilungkiang was able, in the 1960's, to stockpile 52 tractors which were claimed to be obsolete.[13]

In the 1950's, when the stations were being established and when personnel were being trained and equipment tested, it is unlikely that the annual balance sheet for each station was taken as the sole indicator of the station's management or economic rationality. More likely, it was used as an indicator to set the price to charge the cooperatives for plowing. By the early 1960's, however, profit was taken as one main criterion, as will be described in Chapter 8.

10. Li She-nan, "A Few Problems Regarding the Lowering of Operational Costs of Agricultural Machinery," *Jen-min Jih-pao,* June 30, 1964 (SCMP 3272, p. 1).

11. "Summing-up Report on the National Conference of Tractor Station Masters (Excerpts)," *Chung-kuo Nung-pao,* no. 5 (Nov.), 1958 (ECMM 147, p. 19).

12. Peking Agricultural Machinery College, East Is Red Commune, Criticism and Repudiation Office, "P'eng Chen Counter-Revolutionary Revisionist Clique's Crime is Most Heinous," *Nun-yeh Chi-hsieh Chi-shu,* no. 6 (Sept. 18), 1967 (SCMM 610, p. 8).

13. Peking Agricultural Machinery College, East Is Red Commune, Criticism and Repudiation Office, "Completely Settle the Heinous Crimes of China's Khrushchev and Company in Undermining Agricultural Mechanization," *Nung-yeh Chi-hsieh Chi-shu,* no. 5 (Aug. 8), 1967 (SCMM 610, p. 29).

Regardless of the precise significance attached to the profit and loss statements during the 1950's, it seems clear that these data were gathered with great energy, as the figures in Table 4.3 indicate. By 1957, the MTS system was beginning to look economically sensible. Operating costs had been reduced to ¥1.30/mou (15 mou = 1 ha.), and overall losses were down from 21 per cent to 6 per cent of the operating budget. Almost half of the stations were showing a profit (see Table 4.3). It is not clear whether these figures include depreciation, which is far from negligible. If one assumes that a standard tractor cost ¥10,000, can service 100 ha. per year, and lasts 10 years, then depreciation costs ¥0.66 per mou.

Three factors were most important for a station to show profitable operations:[14] The first is the total work done by each tractor. To take into account the obvious differences in capacity between large and small tractors, this factor was always reported in the area serviced for each standard tractor, i.e., for each 15 h.p. The second factor is the fuel used per area serviced. This ratio was used to measure the management efficiency of the station from several points of view. It would reflect on the station's ability to keep machinery in good operating condition; to develop proper field techniques that would minimize fuel wasted in turning or crossing already plowed areas; to select the proper machine for the required task; and to arrange the work efficiently to minimize fuel wasted in getting to the field and back to the station. This criterion seems to have been considered important, and many stations reported a variety of steps to reduce fuel used per mou serviced. Considering China's shortage of petroleum during the 1950's, a high priority attached to this criterion is not surprising.

While both of these criteria were sensible, they generated at least one problem. Neither would reward tractor stations for performing subsidiary work, such as transportation, irrigation, or processing of grains. Thus, tractors tended not to be fully utilized during most of the year. The solution apparently was to establish a method of converting these subsidiary activities into an equivalent for plowing, and then add the equivalent to the basic report.[15]

14. In the Soviet Union during the 1930's pretty much the same criteria were used. Naum Jasny, *The Socialized Agriculture of the USSR* (Stanford: Stanford University Press, 1949), p. 287.

15. "P'eng Chen Counter-Revolutionary Revisionist Clique's Crime," SCMM 610, p. 8.

Table 4.3. Financial accounts of machine tractor stations, 1953–57

Year	Stations[a]	Tractors[b]	Ha. serviced (1000)	Ha./ tractor	Average operating expenses		Total operating budget (¥1000)	Losses (¥1000)[a]	Losses as % of budget	% of stations showing profit[a]
					per mou[a]	per ha.				
1953	9	68	(6)[c]	(87)[d]	2.74	41.2	(240)[c]			0
1954	87	778	(73)[c]	(93)[d]	1.85	27.8	(2,030)[c]	7,290	(21)[c]	8
1955	129	2,203	226[b]	103[b]	1.50	22.5	5,100[e]			25.6
1956	312	9,862	1,350[b]	150[b]	1.39	20.8	27,800[e]			22.1
1957	366	12,036	1,740[b]	145[b]	1.30	19.5	33,800[e]	2,030	6[e]	43.7

Sources:

a. "Summing-up Report on the National Conference of Tractor Station Masters (Excerpts)," *Chung-kuo Nung-pao,* no. 5 (Nov.), 1958 (ECMM 147, pp. 17–18). It is not clear why this article refers to a different number of tractor stations than is shown in Table 4.1.

b. Table 4.1.

c. Computed, based on estimates.

d. Estimated.

e. Computed.

Note: Figures in parentheses are estimates.

A third criterion to which the administrative bureaucracy paid attention was the number of personnel attached to a station. Because wages and fringe benefits were high, there was a tendency for the number of personnel to grow, increasing expenses. At the end of 1957, the machine tractor system had 31,762 employees (office and field) for 366 stations, which possessed 12,036 standard tractors. This was an average of 2.64 workers per standard tractor. A tractor station management conference recommended that the number of workers per standard tractor should be 2.2 to 2.7, the lower number for the larger stations.[16]

Central leaders recognized that profit was not an adequate guide to the performance of stations, and other criteria were established. It is likely that the precise formulation of these criteria differed according to province and year. One formulation, in use by some stations during 1955, was the "three guarantee, four fix system,"[17] under which the stations guaranteed three items: tasks, costs, and sectors of work; and were assigned four fixed standards to achieve these goals: quality, volume of work, consumption of fuel, and number of personnel. In Liaoning in 1955, the criteria mentioned were: (1) handling of more grain, (2) reduction of expenditures, (3) industrious and thrifty operation, and (4) establishment of close links with the masses.[18]

A more comprehensive managerial system was proposed for Liaoning in 1963. First, a long-term plan was made to specify and stabilize the number of tractors, farm implements, personnel, and service stations. Then, to monitor expenses carefully, each tractor team was required to plan and to be responsible for underwriting the expenses for fuel, repairs, and other tasks. It was presumed that such rigorous planning and management would guarantee the quality of work, safety of personnel, and longevity of machines.[19]

Despite all these systems, it appears, judging from the reports of tractor station management conferences, that the greatest emphasis was placed on profit, utilization rates, and fuel economy. These, of course, are the factors that were most easily quantified. The emphasis on quantification had significant implications for management, which will be described below.

16. "Summing-up Report," ECMM 147, p. 20.

17. "Good Results Achieved in the Construction of Agricultural Machine and Tractor Stations in the Country," NCNA Peking, Jan. 10, 1956 (SCMP 1217, p. 14).

18. "Liaoning Calls Conference of State Farms and Other Agricultural Enterprises," NCNA Shenyang, Aug. 15, 1955 (SCMP 1113, p. 23).

19. Chang, "On the Question of Strengthening," SCMM 400, p. 23.

Supplementing the government administration was Party leadership. The MTS's, during the 1950's received leadership from the county Party committee. From 1958 to 1961, leadership was transferred to commune Party committees,[20] but when agricultural machinery stations were re-established after 1962, leadership of the county Party committee was reasserted. A *People's Daily* editorial in November 1963 stated: "The county CCP committee and the county people's council should conscientiously strengthen their leadership over tractor stations in the county."[21]

The particular task of the county Party committee with regard to agricultural mechanization was to integrate the activities of the tractor stations with the production needs of the surrounding cooperatives (or communes). It might organize the stations to send machinery to the cooperatives to demonstrate the potential of the machines and to explain the contract system.[22] After contracts were signed, the committee would see that they were kept, would inspect the work of stations, and would resolve disputes. In one reported case the county Party committee discovered that production teams were being overcharged, and convinced the tractor station, "after heated discussion," that the team should receive a refund.[23] The Party leadership was also concerned with improving efficiency, i.e., utilization of tractors, economy, and profit. Workers at one tractor station summed up the objectives of the county Party committee: "The Party committee above [i.e., at higher levels] wants us to raise our work efficiency and lower cost continuously."[24] The committee had only an indirect influence on the careers of the station personnel; by assisting it in linking up with surrounding cooperatives, it could help the station reach a higher utilization rate for the machinery, and this would be a useful record for a station manager seeking a rising career.

Internal Management

Internally, the machine tractor stations were organized as factories rather than as rural production units. The employees received fixed monthly salaries and food rations as industrial workers did. Some MTS

20. "Suggestions Concerning Strengthening the Management of Agricultural Machinery by People's Communes, Revised Draft," *Nung-yeh Chi-hsieh,* no. 1, 1959, p. 20.

21. "Run Tractor Stations Truly Well," *Jen-min Jih-pao* editorial, Nov. 26, 1963 (SCMP 3115, p. 10).

22. Interview with agricultural machinery technician who worked in China.

23. Chiang Wen-hsien, "How the Tractor Station of T'aiku Hsien Serves Agricultural Production," *Ching-chi Yen-chiu,* no. 4 (April), 1965 (SCMM 472, p. 18).

24. *Ibid.,* p. 17.

employees had a pension plan for retirement and a health plan, both presumably paid for by the state.[25]

Because of the seasonal nature of the work, there was debate about whether MTS employees should receive salaries year round. A management conference in January 1958 recommended that "tractor drivers are to be paid wages by the tractor stations for the period of operation; they are to return to agricultural cooperatives to perform production during the non-operation period."[26] In Heilungkiang, it was reported during the Cultural Revolution that stations had employed temporary workers who received a higher salary than the permanent workers.[27] (Presumably it was a higher monthly salary because it was for only part of the year, and perhaps indicated they did not receive fringe benefits.)

Employees of the MTS system also had a form of job security or unemployment insurance. In 1957/58, when it was decided to cut the personnel of the MTS system to reduce redundancy and waste, two steps were taken to assure that employees would not suffer: present workers would have first claim to new jobs established, and those employees who were not needed and who were sent home to their villages to join the cooperatives would continue to receive wages for a while.[28]

Probably the most important feature of working in the MTS system was the status it conferred. Station supervisors and technicians were national cadre, and in addition were working in the vanguard of agricultural technical transformation, an exciting field with particular prestige.[29]

Little is known about patterns of authority within the tractor stations. Presumably the options—one-man director, Party leadership, or committee leadership—were similar to those throughout industry, and presumably the solutions were about the same.[30]

25. Revolutionary Committee for Jui-ch'eng County, Shansi, "Use Mao Tse-tung's Thought to Direct the Work of Sending down the Farm Machines and Tools of State Stations," *Nung-yeh Chi-hsieh Chi-shu,* no. 11 (Nov. 8), 1968 (SCMM 643, pp. 4, 6).

26. "Summing-up Report," ECMM 147, p. 20.

27. Take-over Committee of the Agricultural Machine Station, Nunchiang County, Heilungkiang, "Thoroughly Eliminate the Pernicious Influence of Material Incentive," *Nung-yeh Chi-hsieh Chi-shu,* no. 4 (July), 1967 (SCMM 605, p. 30).

28. "Summing-up Report," ECMM 147, p. 20.

29. Interview with technician who worked in agricultural mechanization.

30. Excellent analyses of the management issues in the industrial sector are Franz Schurmann, *Ideology and Organization in Communist China* (Berkeley: University of California Press, 1968), pp. 220–308; Barry Richman, *Industrial Society in Communist China* (New York: Vintage, 1969); Charles Bettelheim, *Cultural Revolution and Industrial Organization in China* (New York: Montly Review Press, 1974).

Machine tractor stations and agricultural machine stations had Communist Party branch organizations, as did every other state organ. The Party paralleled the regular administration within the unit, and made all important decisions. Symbolic of its power, the Party headquarters room typically would have the only telephone and the only firearms at the station. The MTS Party branch probably reported to the county Party committee.[31]

When the MTS's were dissolved and tractor stations were put under the direct supervision of communes in 1958, the station Party branch came under the leadership of the commune Party committee. After 1961, however, it was again under the leadership of the county committee.

Special directives apparently went out in about 1961 to strengthen political leadership in the stations. In 1961 the Shantung Young Communist League made plans to establish branches or groups at all tractor stations and in all tractor teams.[32] In Heilungkiang in 1964 there was an instruction from the Central Committee and the provincial committee to strengthen political work on the agricultural machinery front. In response, 362 political bureaus and offices were established in most county-level AMS's and in most branch stations.[33] A similar program was developed in Shensi at the same time.[34]

In Liaoning, which was chosen by the Central Committee as a focal point for developing agricultural mechanization in the early 1960's, the role of the Party organization within the tractor station was spelled out clearly:

> A Party branch (general branch) attached to a tractor station is the basic-level organization of the Party in the tractor station, and is the core of leadership for all kinds of work in the tractor station. The obligations of leadership exercised by the Party organization attached to a tractor station in production and administrative work are: to implement and carry out the guidelines and policies of the Party; to

31. Interview with technician who worked in agricultural mechanization.

32. "Shantung YCL Provincial Committee Issues Notice on Serious Strengthening of YCL Work on Tractor Stations"; "Shantung YCL Provincial Committee Calls upon Young Tractor Drivers in the Whole Province to Plow More Land, Carry Out Plowing Properly, Take Good Care of the Tractors, and Save Fuel"; and Chang Chi-wu, "YCL Organization of the Tractor Station of the Tangkou Commune Gives Young Workers Education Aimed at Raising Their Thinking and Improving Their Technique—Serve Agriculture Whole Heartedly and Give Full Play to the Power of Tractors," all in *Chung-kuo Ch'ing-nien Pao,* Sept. 24, 1961 (SCMP 2603, pp. 10, 12 and 13, respectively).

33. "Heilungkiang Establishes Political Organs in Farm Machinery System," *Nung-yeh Chi-hsieh Chi-shu,* no. 2 (Feb.), 1966 (JPRS 34, 918, pp. 17–18).

34. Li, "A Few Problems," SCMP 3272, p. 1.

insure the complete fulfillment and overfulfillment of the production plans and to realize the different tasks handed down by the supervising organ at a higher level; to discuss and determine all kinds of important problems in work for the tractor station; to exercise supervision over the administrative and leading personnel and find out how they implement and carry out the production plans, the instructions of a higher level, and the resolutions of the Party committee attached to the tractor station.

As a basic accounting unit, the tractor station must enforce the system of making the station master assume responsibility under the leadership of the Party committee. The station master is responsible for the direction of production and administrative work. He must seriously organize the implementation of the decisions of the Party committee to insure that the decisions of the Party committee are brought into realization. The leadership system of integrating collective leadership with personal responsibility must be strictly implemented.

The Party organization attached to a tractor station must tighten organic life according to the provisions of the Party Constitution, strengthen the ideological construction and organic construction of the Party, and bring the role of the Party members into full play.

The Party organization attached to a tractor station must exercise stronger leadership over the trade union and the Young Communist League, bring these organizations into full play, and make them really function as lieutenants to the Party and links with the masses.[35]

Political workers in the stations were often particularly concerned about the relationship of the station with the communes it served and paid special attention to timeliness and quality of plowing. A Party branch in a station might organize a conference of tractor team heads to emphasize the significance of these factors. The Party branch was also concerned about the work style of station personnel, and would hold meetings and conduct education to improve it. Specifically, the Party branch would educate tractor station workers to eat, live, and work with the commune members when they went to a village. They were expected to pay for food and hand in ration coupons for grain they ate. They had to return all borrowed objects and clean up their living quarters. They had to measure accurately the land that they plowed. The Party branch also organized political and ideological study, including study of the thought of Mao Tse-tung.[36] A good way for the Party branch secretary to learn about the operations of the station and the needs of the communes it served was to participate in productive labor and go out with the tractor occasionally on plowing jobs.[37]

35. Chang, "On the Question of Strengthening," SCMM 400, p. 25.
36. Chiang, "Tractor Station of T'aiku Hsien," SCMM 472, pp. 15–24.
37. Chang, "On the Question of Strengthening," SCMM 400, p. 25.

The evidence presented below is that this Party system alone could not achieve the efficient integration of the station with the surrounding farms. In the absence of bureaucratic restructuring, the Party tended to assume the same interests as the administrative bureaucracy and encouraged the efficient completion of the station's plan for plowing. The system of dual leadership—Party and government—did not solve the problems of bureaucratic structure.

MTS's and Agricultural Production

Many factors influence the profitability of mechanization, but perhaps most crucial is the cost of machine services. The price of plowing services appears to have been set by provincial price bureaus, using the balance sheets of the tractor stations to ascertain basic cost. The province could give whatever subsidies were desired.[38] The stations were not free to set a price based on their own costs. In Kwangtung in 1965, for example, the provincial government dropped the standard plowing charge from ¥1.39/mou to ¥1.00/mou.[39] In Shensi in 1964, the fee was ¥1.32; the actual operational costs ranged from ¥3.00/mou (the highest cost), to ¥1.30/mou at an advanced AMS, to ¥0.60/mou at the most advanced AMS. The provincial average was ¥1.80/mou.[40] The standard fee of ¥1.32 would permit the provincial AMS system, after it was running efficiently, to show a slight profit.

Whether the commune could profitably use the labor and animal power released by mechanization was determined by the local economic and political situation. If suitable alternative activities for labor existed—such as intensified cultivation, more multiple cropping, or development of subsidiary economic activities—and if the entrepreneurial skill and political will existed to redirect resources to such activities, then mechanization could add more to the local economy. It should be noted that the tractor station would have virtually no influence over whether the cooperative or commune management committee could locate such profitable outlets and mobilize resources.[41]

38. Evidence that prices were set by the provincial price bureaus is for the 1960's, and it is possible that prices were established nationally, perhaps with regional variations, by the central government in the 1950's.

39. "Prices of Means of Agricultural Production Drop in Kwangtung," *Nan-fang Jih-pao*. Dec. 10, 1965 (SCMP 3609, p. 12).

40. Li, "A Few Problems," SCMP 3272, p. 1.

41. This is subject to some qualification. Only if plowing services are timely and reliable will production units give up their draft animals.

One way the cooperative or commune could benefit from mechanization was increased yields. Because mechanized plowing was more uniform and deeper, yields were generally higher, as indicated in a survey summarized in Table 4.4. At these levels of increase shown, there would be little question of the profitability of machinery. An extra 300 kg/ha. would be the equivalent of 40 catties/mou. If the grain were sold for ¥10 per 100 catties (roughly correct) then the 40 catties would be worth ¥4.00. Since tractor plowing of one mou cost somewhere between ¥1.00 and ¥1.38, there would be an extra profit of ¥2.6 to ¥3.0 for every mou that is machine plowed, plus whatever extra was produced by the manpower and animal power that was released.[42]

Table 4.4. Impact of machine plowing on yield, Heilungkiang, 1953–56

Year	Land investigated (ha.)	Yield, machine-cultivated land (Kg./ha.)	Yield, animal-cultivated land (kg./ha.)	Increase	% of increase
1953	198	1,347	1,165	182	16
1954	1,316	1,827	1,375	452	33
1955	3,002	1,871	1,497	374	20
1956	7,529	1,983	1,594	389	24

Source: Wang Kuang-wei, "On the Modernization of China's Agriculture," *Hsüeh-hsi*, no. 1 (Jan.), 1958 (ECMM 128, p. 34). The precise character of these experiments is not clear from the report. For example, it is not clear whether other inputs—fertilizer and water—are controlled for, or whether the same crops are involved.

While mechanization has the potential to increase yield and/or cropping index, the increase in yield is never automatic. Nature imposes strict requirments upon plowing, as upon other aspects of mechanization: (1) The plowing must be done at the appropriate time in the agricultural season; delay of even a few days can cause problems with sowing or harvesting. It must be reliable so that animals can be reassigned. (2) It must be uniformly deep. (3) It should respect the natural contour lines. (4) It must be done in a manner consistent with local cultivation practices such as ridge farming or intercropping.

42. The concept of profitability is restricted to the idea that the agricultural production unit would show a profit in its accounting procedure. Whether it would, in fact, be profitable for the national economy would depend on the rationality of the prices for grain and for plowing, factors beyond this discussion.

As a practical matter, there have been substantial difficulties in meeting these requirements because many of the regulations, evaluation criteria, incentives, etc., enumerated above did not adequately stress these factors. One of the most important criteria to MTS personnel was fuel economy, and to achieve it they took various steps that harmed agricultural production. The most common was failure to plow to the specified depth. During the Cultural Revolution this problem was explained: "Some agricultural machine stations of Peking . . . use the methods of carrying out shallow plowing at a high speed, plowing land to a greater depth at the head of fields but carrying out shallow plowing in the middle of fields; carrying out deep plowing when there are persons to make inspection but shallow plowing when no inspection is carried out to hoodwink the peasants.[43] Sometimes tractor drivers reduced depth when working at night, when inspection was difficult.[44] A county CCP committee in Shensi conducted an investigation of tractor services in 1964, and reported that "in some fields the quality of plowing was not good, and the masses were displeased with this. It was also discovered that, in exceptional cases, tractor teams reported shallow plowing as deep plowing or overstated the acreage plowed with a view to over-charging."[45] Cultural Revolution critiques link such practices directly to the institutional arrangements of the MTS and AMS. "Some tractor drivers only considered the question of plowing a greater area of land and getting more cash rewards and did not consider the needs of agricultural production and did not pay close attention to the quality of their work."[46]

Timing presented another problem.[47] There were several reasons for delays. First was the problem of reliability of the machines. Spare parts were in short supply, and the situation was complicated by the fact that China's tractor stock came from many countries and by the lack of trained operators and repairmen.

43. "P'eng Chen Counter-Revolutionary Revisionist Clique's Crime," SCMM 610, p. 8.

44. "It is Good for Tractors to Revert to Chairman Mao's Revolutionary Line—Report on Investigation of Change in Management of Tractors by the Collective in Lank'ou County, Honan," *Nung-yeh Chi-hsieh Chi-shu,* no. 10, (Oct. 8), 1968 (SCMM 643, p. 22).

45. Chiang, "Tractor Station of T'aiku Hsien," SCMM, p. 18.

46. "It is Good for Tractors," SCMM 643, p. 22.

47. In the Soviet Union a dual rate structure was established in 1947 to give the MTS's an incentive to perform operations on time. Late operations cost roughly 20–25 per cent less than timely ones. Jasny, *Socialized Agriculture,* p. 292; Laird, Sharp, and Sturtevant, *The Rise and Fall,* p. 57. There is no indication that this was used in China.

A second reason for delays was that the tractor stations wanted to get all the orders for plowing before beginning so that the tractors could be dispatched in the most efficient, economical manner. By the time the orders were all in and the tractor routes planned, some of the production teams might have decided it was too late and begun plowing with animals. The problem of timeliness was so serious that cadres of one brigade summed up the situation in these words: "Tractors of the state-operated station do not come when you need them; yet they are here when you don't need them."[48]

Another problem concerned tractor services to poor, outlying production teams. They had difficulty in paying in advance, so the tractor stations tended to overlook them. Moreover, since they were far away, more fuel would be expended getting to them. If, in addition, they were in mountainous areas with small plots, additional inefficiencies would be encountered. Thus the stations tended to refuse to service such areas.[49]

Corruption was also a problem. There were reports that bribery was required to get decent service from machine operators. In one location peasants reported that "the tractor drivers worked better if they were well treated. They would not work so well if they were not treated so well. If they were given good food, the land was plowed to a great depth."[50] Some commune members composed a song:

> If there is food but no wine, the tractor does not move;
> If there is wine and no food, the tractor runs out of order;
> If there are both wine and food, the tractor flies.[51]

Finally, there were difficulties in financial relationships. The stations sometimes had trouble collecting fees from the production units. One tractor was owed ¥370,000 in fees from the surrounding communes.[52] The tractor station, reasonably enough, began to demand cash payment in advance, upon signing the contract with the production unit: "When a production team wanted a tractor to plow its land it had to send somebody to take money to the tractor station in the county seat to sign a contract, in which were stated the parties concerned and the organ supervising the work to be done, and which we marked with big and small seals in ample

48. "It is Good for Tractors," SCMM 643, p. 18.
49. *Ibid.*, p. 19; "P'eng Chen Counter-Revolutionary Revisionist Clique's Crime," SCMM 610, p. 8.
50. "It is Good for Tractors," SCMM 643, p. 22.
51. *Ibid.*
52. *Ibid.*, p. 20.

space."[53] The result of this legalistic approach? "It often happened that the farming season was already over when the contract was signed and the tractor came to do the job."[54] In addition, this emphasis on cash payment resulted in certain losses in fuel economy, apparently judged less important in these circumstances by the station management:

> They plowed the land first for the production team that paid the money first. They basically did not consider which production team needed a tractor more urgently. Some production teams had relatively serious conditions of sand and salt in the soil, had insufficient draft animals and were in relatively poor financial situation. They therefore needed the work of machines more urgently. But it often happened that when they had raised enough money and hurried to the tractor station to sign the contract, the tractors were already sent to other communes.[55]

(It was not explained why there was no rural credit available to these production teams.)

The net result of these problems was an overall failure to integrate machinery fully and efficiently into agricultural production.[56] General contradictions and inefficiencies characterized the system: "It is not easy to link the mechanized plans of the state-owned stations with the production plans of communes and teams. It is not easy to make over-all arrangements for and effect the close coordination of mechanical power, animal power and manpower."[57] One cooperative in Heilungkiang, Hsingyeh (which was later the site of decentralization experiments to be discussed below), described the problems in coordinating machines with the farming schedule: "Lack of harmony frequently occurred in the process of production. When machines were needed, they could not be made available at once, and when they were finally sent over, they sometimes could not be usefully employed. This greatly harmed production."[58] Another

53. *Ibid.*, p. 18. Similar problems are described in "P'eng Chen Counter-Revolutionary Revisionist Clique's Crime," SCMM 610, p. 8.

54. "It is Good for Tractors," SCMM 643, p. 18.

55. *Ibid.*, pp. 18–19.

56. Pretty much the same types of problems were observed in the Soviet Union. Jasny, *Socialized Agriculture*, p. 288; Volin, *A Century of Russian Agriculture* (Cambridge: Harvard University Press, 1970), p. 454.

57. Liao Nung-ko, "Knock Down China's Khrushchev and Completely Discredit by Criticism His Line of State Monopoly," *Nung-yeh Chi-hsieh Chi-shu*, no. 2, 1968 (SCMM 620, p. 8).

58. Work Group of CCP Nunkiang District Committee of Heilungkiang, CCP Paichuan County Committee of Heilungkiang, and Heilungkiang Branch of the New China News Agency, "The Universal Establishment of People's Communes Accelerates Mechanization—An Investigation into Hsingnung People's Commune in Paichuan Hsien, Heilungkiang Province," *Jen-min Jih-pao*, Dec. 4, 1959 (SCMP 2160, p. 10).

report from the same location specified in greater detail the variety of factors that inhibited efficient utilization of agricultural machinery.

> While technical force was concentrated to facilitate control and repair, the working plans were not flexible because land holdings were scattered, and it frequently happened that the tractor stations could not meet the requirements of the cooperatives in time. Meanwhile, the frequent movement of engines over long distances caused much waste of working time. Moreover, the cooperatives found it difficult to direct the tractor stations while the tractor operators, who had no direct interests in the results of cooperative production, frequently overlooked the farming quality. Also the fees charged were high. Because of this, the masses did not welcome this form of operation.
>
> . . . for a time, we mechanically applied farming methods suited to the natural conditions of foreign countries and introduced agricultural machinery designed and made according to the conditions of foreign countries. The result was that as such farming methods and agricultural machinery did not entirely suit the local conditions of our country, production either declined or showed no marked increase.[59]

A *People's Daily* editorial, reacting to a conference on agricultural machinery management in the summer of 1965, summed up the institutional contradiction:

> Of course, due to the different systems of ownership practiced by the farm machinery station and the commune, each keeps its own accounts. In some cases, as a result of the unsatisfactory arrangements made by both sides, contradictions such as "two accounts" and "two hearts" have come to the fore. In these circumstances, machinery stations, being only concerned about their own profits and losses, tend to disregard the increased output and increased income of communes and production teams. The latter, on the other hand, may concern themselves only in the use of farm machinery without caring for the profit or losses of farm production and agricultural mechanization.[60]

Considering all these problems with the MTS system, it is not surprising that many peasants were less than enthusiastic about agricultural machinery. Even in Heilungkiang in 1965, where agricultural mechanization was well developed, it was admitted that "the peasants still have doubts over the use of machinery. First they fear that machinery will not increase output and increase income. Second they fear that the quality of work

59. Li Chien-pai, "Questions of Agricultural Mechanization," *Hsüeh-hsi,* no. 10–11 (May 31), 1958 (ECMM 140, p. 37).

60. "Management of Agricultural Machinery Should Better Serve Agricultural Production," *Jen-min Jih-pao* editorial, Aug. 31, 1965 (SCMP 3543, p. 11).

will not be good. And third, they fear that the machinery will bring with it various obstructions, and cause delays in farming time."[61]

Very little is known about the bargaining power of the communes with respect to the MTS's. Agricultural production units had the legal right to refuse to contract with the tractor stations. The sixty-point draft rules and regulations for the communes of March 1961 assured the production team the right to refuse to purchase defective agricultural inputs, and the revised draft of September 1962 gave absolute rights to refuse purchase.[62] With regard to tractor services, however, it seems probable that a county Party committee would strongly urge a commune or team to purchase tractor services if possible, to increase the utilization rate of the tractors.

If a commune or team were dissatisfied with the quality and/or timing of its plowing, probably its best recourse would have been to the county Party committee. Some production teams refused to pay their bills, perhaps as a protest. The regular courts would not intervene in such a dispute, unless corruption were involved.

Of course, we have no way of knowing how widespread the problems described above actually were. Most likely, the examples cited are cases of unusually poor relations between the tractor stations and the surrounding areas. The actual relationship would depend on a variety of factors, including the basic economic logic of mechanization in the area, practice and experience in organizing the machinery, and the sensitivity of county Party leadership to potential problems.

Central authorities were constantly concerned about the ability of the tractor stations to provide suitable services to the communes. If mechanization proved unprofitable, the communes would not purchase services from the stations and the stations would lose more money. Equally important, the collective pattern of economic management would not be reinforced by mechanization. Thus some central authorities were very concerned about quality of service,[63] but the evidence suggests that the administrative control system could not adequately deal with the question of quality, partially because it could not be quantified. The Party control system, which was charged with the responsibility of improving quality,

61. Chao Chen-hua, "Heilungkiang Agricultural Machinery Station Sees Year of Great Change," *Chung-kuo Nung-yeh Chi-hsieh,* no. 6 (June), 1965, pp. 13–16 (JPRS 33, 727, pp. 3–4).

62. Union Research Institute, *Documents of Chinese Communist Party Central Committee,* Vol. 1 (Hong Kong: Union Research Institute, 1971), pp. 695–725.

63. Chang, "On the Question of Strengthening," SCMM 400, p. 22.

was not able to do this, partially because it tended to emphasize meeting production plans. In short, despite the central authorities' awareness of the importance of quality of machine service, the state-owned tractor station system had great difficulties in serving agricultural production efficiently.

Dissatisfaction with the MTS's

As early as 1956, some leaders in China were dissatisfied with the MTS system, and during 1957 there were experiments with alternative patterns of management. These culminated in a massive decentralization program in which agricultural machinery was transferred to communes during 1958 and 1959. Commune ownership of machinery presented its own problems, however, and the state ownership system was again instituted.

One source of dissatisfaction was the specific contradictions and inefficiencies in the MTS system during the mid-1950's, as described above. Complaints were not publicized (as they were for the state mechanized farm system) at a conference to review the MTS system's work in 1955.[64] At the end of 1955, the public mood was one of confidence, and energies seemed devoted to expansion rather than to change. Published reports on the work of 1956 were sparse and did not indicate that serious management problems had emerged.[65] Nevertheless the inefficiencies of the system left many uneasy.

The concrete problems merged with broader issues of overall rural policy with the development of people's communes in 1958. During 1956 and 1957, the agricultural producer cooperatives had expanded in size and function, and confusion emerged concerning the respective roles of the cooperative—theoretically responsible for economic administration—and the county and *hsiang* (township) government. The formation of communes solved this contradiction by merging economic and political

64. "More Tractor Stations," NCNA Peking, Dec. 13, 1955 (SCMP 1190, p. 10); "China Has 139 Tractor Stations," NCNA Peking, Jan. 12, 1956 (SCMP 1209, p. 9); "Kiangsu Plans to Develop Tractor Stations in Next Two Years," NCNA Nanking, Nov. 16, 1955 (SCMP 1179, p. 18); "Hopei Builds More Tractor Stations," NCNA Peking, Jan. 8, 1956 (SCMP 1205, p. 20); "More Tractor Stations for Northeast Province," NCNA Shenyang, Jan. 7, 1956 (SCMP 1205, p. 21); "Liaoning Calls Conference of State Farms and Other Agricultural Enterprises," NCNA Shenyang, Aug. 15, 1955 (SCMP 1113, p. 23).

65. "325 Tractor Stations in China," NCNA Peking, Feb. 26, 1957 (SCMP 1480, p. 10).

administration.[66] Putting agricultural machinery under the control of the commune was a natural consequence of the formation of communes.

This was not simply an administrative question; it involved broad political issues with crucial international dimensions. Decentralization was part of a general dissatisfaction with the Soviet model of development. While the Chinese became aware of the political problems and economic inefficiencies of the Soviet model on the basis of their own experience, Khrushchev's speech to the Twentieth Congress of the Communist Party of the Soviet Union in February 1956 made it publicly obvious that there were serious weaknesses in the Soviet system. While the Chinese may have been miffed by Khrushchev's failure to consult them before the speech, they did not ignore the substance of the critique of the system.

Mao and other Chinese leaders were quick to see problems in the Soviet style of centralized bureaucratic organization. In April 1956, at a meeting of the CCP Politbureau, Mao observed:

> Only by changing the system can we change the way of doing things, and we must give some power to those below. Our discipline has come mostly from the Soviet Union. By being too strict we will tie the hands of the people. If we cannot crush bureaucracy in this way, then the proletarian dictatorship must have an appropriate system. . . . On May 1 . . . May 1 slogans of the Soviet Union need not be posted.[67]

The protests in Poland in October 1956 and the rebellion in Hungary in October–November 1956 confirmed that widespread dissatisfaction and violent resistance to systems modeled after the Soviet Union's were possible. Mao cautioned against adopting the Soviet system indiscriminately and mechanically:

> Was the Hungarian incident good or bad? Problems should be exposed. It is good to expose them. A pustule will burst and emit pus. Those countries that imitated the Soviet pattern and failed to face up to reality have not been doing well and, in fact, have gone wrong. Herein we learn a lesson. What we do should be based upon the universal truths of Marxism-Leninism and should be in accordance with reality in China. We have brought up the slogan "Learn from Soviet experience," but never have we proposed to learn from the Russians' backward experience. Do they have any backward experiences? Yes, they do.
>
> We are followers of Marxism but we do not blindly copy Soviet experience. To do so would be a mistake. Our industrial and commercial transformation and agri-

66. A. Doak Barnett, *Cadres, Bureaucracy, and Political Power* (New York: Columbia University Press, 1967), p. 324.

67. Mao's speech at Expanded Meeting of CCP Political Bureau, April 1956, *Miscellany*, p. 30.

cultural cooperativization differ with the Soviets'. In the several years after agricultural cooperativization, their production decreased, but our production has increased.[68]

The following month, Mao expressed the same idea:

We still should learn from the Soviet Union. We can learn a lot of things from them. But we should learn selectively. We should learn progressive and useful things and should study their blunders critically. . . . Knowledge should be sought everywhere. It is too monotonous to seek it only in one place.[69]

One aspect of the critique of Stalinism and the scrutiny given to the Soviet model was a widespread re-evaluation of the MTS system in the Soviet Union and elsewhere, and the Chinese were well informed of these debates. In the spring of 1956 (March 17–April 9), K'ang Sheng led the Chinese delegation to the Third Party Congress of the Socialist Unity Party in East Germany. In traveling to and from Berlin, he presumably spent a few days in the Soviet Union.[70] The MTS was discussed extensively at the congress in East Germany. It was criticized for forcing consolidation of privately owned land into large plots and for failure to maintain machinery properly. Plans were discussed to evaluate the stations' performance not according to the simple area plowed but rather according to the growth in agricultural production of the land serviced and the growth in the value of the work day for the cooperatives serviced.[71]

K'ang Sheng could have gotten more ideas about the MTS system in the Soviet Union, since he was there at precisely the moment when the system was being reconsidered. A *Pravda* editorial of March 26, 1956, revealed some of the problems: "It is considered essential to relate the pay of MTS directors and specialists to the quantity of produce obtained by the collective farms within an MTS zone and to the fulfillment of

68. Mao's Instructions at a Discussion Meeting Attended by Some of the Delegates to the Second Session of the First Committee of the All-China Federation of Industry and Commerce, Dec. 8, 1956, *ibid.*, pp. 40, 38.

69. Mao's Interjections at a Conference of Provincial and Municipal Committee Secretaries, Jan. 1957, *ibid.*, p. 57.

70. Donald Klein and Anne Clark, *Biographic Dictionary of Chinese Communism* (Cambridge: Harvard University Press, 1971), p. 427. Klein has indicated that the normal pattern for Chinese leaders traveling to Eastern Europe was to spend a few days in Moscow both coming and going.

71. This account is based on the report on agriculture presented to the congress, available in *Tret'ye Konferentsiya Sotsialisticheskoi Edinoi Partii Germanii, 24–30 March 1956* (Moscow: Political Literature Publishers, 1956), pp. 105–110. I am indebted to Toby Trister, of Columbia University's Research Institute on Communist Affairs, for assistance in translating this Russian source.

procurement and purchase assignments, in order to raise the directors' and specialists' material interest in their work.''[72] While this editorial postulated the continuance of the system, more radical rumblings existed in the USSR. In 1955 in some locations a few tractor stations were attached exclusively to specific collectives. Field operations were merged, although separate accounts were retained.[73] Then, in February 1956, two kolkhoz chairmen in the Moscow region asked to purchase tractors. The Soviet MTS system was finally dissolved in 1958.

The MTS system as practiced in the Soviet Union must have made a strong negative impression on K'ang Sheng, who may have been predisposed to criticism.[74] When he returned, he surveyed China's MTS system in depth.[75] He spoke about agricultural mechanization in November 1956. (It is not known whether K'ang's speech was a report to the Second Plenum of the Eighth Central Committee, which met November 10–15, 1956.) Putting together various quoted fragments of K'ang's speech, its main thrust can be reconstructed:

> Collective farms in the Soviet Union have many machines, but the farm output is very low and the cost very high.[76]
>
> How to take the mass line and not divorce ourselves from the masses in introducing mechanization of farming is the most fundamental question. It won't do to belittle science, nor will it do to go without a mass line.[77]
>
> In short, this problem must be solved: how to link the tractors to the peasants. Run in their present form it is definite that tractor stations will develop greater and greater contradictions and will be detached from the masses. If we cannot operate them well, they would become in a disguised form revenue-collection organs, or organs practicing blackmail. The Soviet tractor stations are blackmailers.[78]

72. *Current Digest of the Soviet Press* 8:13 (1956), 30.

73. Miller, *Hundred Thousand Tractors,* pp. 318–319.

74. P. P. Vladimirov, Comintern liaison officer and TASS war correspondent who worked in Yenan from 1942 to 1945, claims that the ''repressive machinery of Kang Sheng, head of the secret services of the Special Area, ruthlessly dealt with those Communists who were friendly towards the Soviet Union'' (*China's Special Area, 1942–1945* [Moscow: APN Publishing House, 1973], summarized in K. Smirnov, ''Exposure of Maoism,'' *Izvestia,* July 22, 1974, Novosti Press Agency no. 30:248, Aug. 1974).

75. ''Completely Settle the Heinous Crimes,'' SCMM 610, p. 28; ''History of Struggle,'' SCMM 633, p. 10.

76. ''Completely Settle the Heinous Crimes,'' SCMM 610, p. 28.

77. ''History of Struggle,'' SCMM 633, p. 13.

78. ''Completely Settle the Heinous Crimes,'' SCMM 610, p. 28; partially quoted in: Eighth Ministry of Machine Building, United Committee of the Revolutionary Rebels, ''57'' United Detachment, ''Wipe out State Monopoly and Promote Mechanization on the

It is definite that agriculture must be mechanized in China, but it is also definite that dogmas cannot be adopted. The cause of agricultural mechanization must be made to establish flesh-and-blood ties with the peasant masses.[79]

In regard to the way of handing over tractors to cooperatives for operation, this must be experimented with and studied.[80]

K'ang was not criticizing poor management in existence at the moment, but was predicting that conflicts of interest between the station and the agricultural production unit would eventually lead to problems unless strong actions were taken. The suggestion to experiment with handing over tractors to cooperatives was, of course, tantamount to suggesting the dissolution of the MTS system. With such a suggestion in the air it is no wonder that the MTS system's review of its work in 1956 was quiet.[81]

(It should be pointed out that interest was expressed in decentralizing many sectors of the Chinese economy in late 1956. Liu Shao-ch'i's speech to the Second National People's Congress in September 1956 showed great interest in decentralization. Experiments were made with different patterns of factory management at the end of 1956 and the beginning of 1957, borrowing to some extent from the experience of other socialist countries also, in this case Yugoslavia.[82] Chinese interest in decentralization paralleled Soviet experimentation with decentralizing economic planning on a regional basis.)

A few years later Mao explained his agreement with K'ang Sheng's analysis. Mao characterized Stalin's agricultural policy by saying that Stalin "did not trust the peasants and did not want to let go of the farm machinery. While maintaining that the means of production belonged to the state on the one hand, on the other hand he felt that the peasants could not afford them. Actually he was only deceiving himself. The state exercised tight control over the masses."[83] Mao indicated that despite collectivization and mechanization, Stalin's agricultural policy was a failure in raising production: "In the year [Stalin] died the [grain] output was the

Basis of Self-Reliance in a Big Way," *Nung-yeh Chi-hsieh Chi-shu*, no. 6 (Sept. 18), 1967 (SCMM 610, p. 10).

79. "Wipe Out State Monopoly," SCMM 610, p. 10. 80. *Ibid.*

81. "325 Tractor Stations in China," NCNA Peking, Feb. 26, 1957 (SCMP 1480, p. 10).

82. Paul Harper, "Workers' Participation in Management in Communist China," paper for American Political Science Conference, 1970, pp. 13–14.

83. Speech on the Book *Economic Problems of Socialism*, Nov. 1959, *Miscellany*, p. 130.

Table 4.5. Experiments with cooperative ownership and management of agricultural machinery, 1957

Experimental cooperative[a]	Location[a]	Ownership	Management	Results or comments
T'ients'un	Peking	State owned machines, collected depreciation charges.[a]	Cooperative managed machinery at all test sites.	T'ients'un and Changkuochuang: (1) Tractor utilization went up from 2–3 months/year to 8 months/year. Field management techniques were changed;[b] (2) Tractors were in use 10 months/year.[c]
Changkuochuang	Peking			
Mincheng	Mishan County, Heilungkiang	At first, state owned machines; leased to cooperatives.[d] Later, cooperatives purchased machines from MTS.[a]		Mincheng: no reports.
Hsiengyeh (became part of Hsingnung Commune)	Paich'uan County, Heilungkiang			Hsiengyeh: (1) Drivers sent to cooperatives earn 30%–40% more; have private plots; dependents take part in production;[e] (2) "A good experience";[f] (3) Yield up 22%, more threshing;[d] (4) Commune spent ¥170,000 for 20 tractors;[g] (5) Larger cooperatives formed in fall 1957 to simplify management and investment;[d] (6) Cost of plowing reduced from ¥1.53 to either ¥0.64 or ¥0.77/mou;[e,h] (7) Utilization up to 7,000–13,000 mou/tractor; quality improved.[e]

Sources:

a. "History of Struggle," SCMM 633, p. 13.

b. Huang Ching, "The Problem of Farm Mechanization in China," *Jen-min Jih-pao,* Oct. 24, 25, 1957 (SCMP 1662, p. 24).

c. Huang Ching, "Simultaneous Development of Industry and Agriculture and the Question of Agricultural Mechanization—Report Delivered at the Meeting of Cadres from Various Organs Directly Subordinate to the CCP Central Committee, Various Organs of the Central Government, the CCP Peking Municipal Committee, and Various Units of the PLA Stationed in Peking," *Hsüeh-hsi,* no. 2 (Jan. 18), 1958 (ECMM 128, p. 54).

d. Heilungkiang, Work Group of CCP Nunkiang District Committee et. al., "The Universal Establishment of People's Communes Accelerates Mechanization, SCMP 2160, p. 10.

e. "Summing-up Report," ECMM 147, p. 18.

f. "Agricultural Mechanization Bears Fruit in Northeastern 3 Provinces," NCNA Harbin, March 18, 1958 (SCMP 1740, p. 8).

g. "Tractors Improve Farming Technique in Northeast China Communes," NCNA Harbin, Sept. 12, 1959 (SCMP 2098, p. 23).

h. Li Chien-nai, "Questions of Agricultural Mechanization," ECMM 140, p. 37.

same as in the Czarist era, and if [Khrushchev] had not changed the policy, the situation would have become increasingly grave."[84]

On the basis of K'ang Sheng's suggestions, experiments were conducted on four cooperatives in 1957 using two forms of tractor ownership/management. Two test sites were in suburban Peking (T'ients'un and Changkuochuang cooperatives) and two in Heilungkiang (Hsingyeh and Mincheng cooperatives).

At the Peking sites and at Hsingyeh for a while, the system tried was "state owned, cooperative managed," in which the cooperatives leased machinery from the tractor station at a fixed rate. The tractor station was responsible for fuel, maintenance, repairs, and staff expenses. The cooperative had complete freedom to assign the tractor. (Results of these experiments are summarized in Table 4.5.)

At Hsingyeh, this system led to a decrease in the cost of plowing from ¥1.53 per mou to ¥1.20.[85] Problems remained, however, centering around fuel and maintenance costs:

> The operators still did not interest themselves in the results of production. The cooperatives, which used but did not own the machines, did not take good care of the machines, causing excessive wear and tear.[86]

> A new contradiction was discovered later. On the one hand, the higher cooperatives always wanted to get more work done with the tractors, and were sometimes liable to be negligent with their maintenance. On the other hand, the tractor drivers wanted to consume less oil. Not enough consideration was therefore given to the quality of work, and production was thus harmed.[87]

To solve this new contradiction, Hsingyeh tried another solution. The cooperative bought the tractors outright from the station. The purchase meant that the cooperative would assume all fuel and maintenance costs. The cooperative also assumed personnel expenses, but was able to do this in a manner that reduced costs. The tractor hands were taken off monthly salary and became members of the cooperative, receiving work points as the peasants did. Their wage income went down, but they could now have private plots and their families could take part in collective labor; it

84. Talk at Symposium of Hsin, Lo, Hsu and Hsin Local Committees. Feb. 1959, *Miscellany*, p. 161.

85. Li Chien-pai, "Questions of Agricultural Mechanization," *Hsüeh-hsi*, no. 10–11 (May 31), 1958 (ECMM 140, p. 37).

86. *Ibid.*

87. Heilungkiang, Work Group of CCP Nunkiang District Committee et al., "The Universal Establishment of People's Communes Accelerates Mechanization," SCMP 2160, p. 10.

was claimed that their total family income went up 30 to 40 per cent. The cost of plowing was reduced to ¥0.64–¥0.77/mou, and tractor utilization went up to 470–870 ha. per tractor. The experiment was considered a great success.

The experiments in Peking, in which the state retained ownership and leased the equipment out, were also reported to be successful. Tractor utilization went up from 2–3 months to 8–10 months because the tractors were used for sowing and transportation in addition to plowing.[88]

It is interesting to note the role of ideology with regard to ownership of agricultural machinery. Li Chien-pai, secretary-general of Heilungkiang's CCP, pointed out that some people considered the form of state ownership (i.e., MTS ownership) to be better from an ideological point of view: "A view was held for a time that the form of cooperative ownership and operation was collective ownership and was not so advanced as the ownership by the State and all the people and that to bring the tractor stations under cooperative ownership and operation was a sort of 're-trogression.' "[89] Li Chien-pai had no use for such an argument:

> In our opinion, this view is one-sided. Both the ownership by all people and the collective ownership are socialist ownership. The form of agricultural mechanization should be selected on the basis of *practical considerations;* that is, the form that can best develop socialist agricultural production and can build socialism more, better, faster and more economically should be selected. One must not indulge in empty talk about advanced forms regardless of reality or select a form according to his subjective desire.[90]

The results and implications of these experiments were spelled out at a meeting of tractor station directors in January 1958.[91] Understandably, the tractor station masters were reluctant to see their stations sold out from under them; the resolution of the meeting reflected their concerns by cautiously suggesting delay. "In the not distant future, the masses are expected to own and manage agricultural machinery themselves. This will be the main form of management. But it should be borne in mind that state-owned tractor stations remain the main form of agricultural mechanization in our country at the present moment."[92] More experimentation

88. Documentation of all figures may be found on Table 4.5.
89. Li, "Questions of Agricultural Mechanization," ECMM 140, p. 37.
90. *Ibid*. Emphasis added.
91. "Tractor Station Conference," NCNA Peking, Feb. 16, 1958 (SCMP 1719, p. 4); "China's Work on Agricultural Mechanized Tractor Stations During Period of First Five-Year Plan Reviewed," NCNA Peking, Feb. 16, 1958 (SCMP 1721, p. 33).
92. "Summing-up Report," ECMM 147, p. 16.

was recommended: ''In those areas where conditions permit, more experiments should be carried out positively so that, it is hoped, in a year (or longer) experiences will be gained in order to create the conditions for the promotion of this form of operation.''[93]

The central organs moved quickly, however. At the Chengtu Conference in March 1958, mechanization policy was reviewed and a comprehensive position paper, ''Views on the Question of Agricultural Mechanization,'' was adopted. This recommended decentralizing ownership of agricultural machinery. The consequences of this decision will be examined in Chapter 6, but first it is useful to examine the background for this statement, because it deals with the broad role of agricultural mechanization after cooperatives had been established. This review will be the focus of the following chapter.

93. ''History of Struggle,'' SCMM 633, p. 13.

5. Strategy of Agricultural Intensification

Once collectivization of agriculture was achieved, a basic institutional question of rural development was substantially resolved. Rural social organization would be based on collective ownership by a group of a few dozen families. Remuneration would be determined primarily by labor and by the profitability of the basic accounting unit. In subsequent years there were some debates on the manner in which the collective economy would be administered and on the balance between the dominant collective economy and the small private economy that remained. These debates came to a peak in 1962 and will be described later.

In a certain sense, the resolution of institutional issues in 1956 placed new demands on agricultural production. Mao was exceedingly sensitive to the importance of gains in agricultural production to consolidate the socialist transformation in agriculture, industry, and commerce. At a meeting in 1957, Mao pointed out:

> We should pay attention to foodgrain production. It will be disastrous if we don't. When we have food, we have everything. If there is food to eat even during a labor strike or a student strike, no great disturbance will occur. . . .
>
> Last year the Soviet Union had a bounteous harvest and it was easy to do a lot of things. With last year's experience we should try to reap a bounteous harvest this year. If we all work hard to get it, it will be of great significance to the world communist movement. In history, the socialist cooperative movement resulted in decreased production. Our increase in production in 1956 was not much. We should strive for a major increase in 1957.[1]

At the same time, Mao emphasized that collectivizing agriculture had not automatically solved production problems. Indeed, collectivization brought with it risks of mismanagement. Mao warned:

1. Mao's Interjections at Conference of Provincial and Municipal Committee Secretaries, Jan. 1957, *Miscellany,* p. 46.

On the question of agriculture the experience of some socialist countries proves that even where agriculture is collectivized, where collectivization is mismanaged it is still not possible to increase production. The root cause of the failure to increase agricultural production in some countries [the Soviet Union?] is that the state's policy towards the peasants is questionable. The peasants' burden of taxation is too heavy while the price of agricultural products is very low, and that of industrial goods is very high.[2]

Fortunately, the methods for increasing agricultural production could now be partially separated from the question of whether and when to collectivize agriculture. No longer was strengthening the collective economy the primary objective for agricultural technology. Overall productivity and social change became equally relevant. During 1957–58, the Chinese leadership concluded that the way to increase agricultural productivity was through a tremendous intensification of cultivation. Labor would be mobilized to tend crops more carefully, to gather natural fertilizers, to manage water, etc. In fact the intensification program was so massive that even China's vast labor supply would prove inadequate at certain seasons, and tool reform would become an integral part of the program.

While there was some logic in this strategy, it required large-scale supplies of modern industrial inputs, such as chemical fertilizer, mechanical pumps, energy, and improved tools and irrigation systems.[3] Without the added supplies, the delicate balance of China's agriculture was disrupted; when this disruption was combined with bad weather in 1959–61, near disaster resulted.

The Twelve-Year Plan for Agricultural Development

Symbolic of this new emphasis on production was the adoption of the Draft Program for Agricultural Development in the People's Republic of China, 1956–1967, on January 23, 1956.[4] The program was a comprehensive statement for rural development. Naturally enough, it started with

2. "On the Ten Great Relationships," April 25, 1956; available in Stuart Schram, ed., *Chairman Mao Talks to the People,* (New York: Pantheon, 1974), p. 64.

3. Mao recognized the needs for these inputs and (prematurely) announced in 1957 that every province had established its own chemical fertilizer plant. Perhaps he was really encouraging them to do so. Mao's Interjections at Conference of Provincial and Municipal Committee Secretaries, Jan. 1957, *Miscellany,* p. 51.

4. The text is conveniently available in Robert Bowie and John Fairbank, *Communist China, 1955–1959* (Cambridge: Harvard University Press, 1965), pp. 119–126. An analysis of the political conflict over this plan is Parris Chang, *Power and Policy in China* (University Park: Pennsylvania State University Press, 1975), pp. 9–46.

the proposal (which was by January 1956 almost a reality) that agriculture be collectivized. It set as a target that 85 per cent of all peasant households should be in agricultural producer cooperatives by the end of 1956. Moreover, it suggested that in regions where cooperatives had a firm foundation, the changeover to the advanced form of cooperatives should be completed in 1957.

The Draft Program touched on many other aspects of economic, social, and cultural development as well. It specified plans for distribution of telephones; it urged reforestation and attention to aquatic products; it asked for equal pay for women; it encouraged young people to learn about science; it noted the need to eliminate urban unemployment and to improve economic and cultural exchanges between the cities and the countryside.

From the point of view of agrotechnical development, the Draft Program called for higher yields through intensification of cultivation. The plan urged extensive expansion of multiple cropping, holding up these targets:

South of Wuling Mountains	230%
Between Wuling Mountains and Yangtze River	200%
Between Yangtze and Yellow rivers, Tsingling Mountains and Pailung River	160%
Between Yellow River and Great Wall	120%

Largely through these increases, higher annual yields were projected. For the vast area south of the Huai River, a goal of 6 tons per hectare per year was set. For the region between the Huai and Yellow Rivers a target of 3.75 tons per hectare was established. In North China, the goal was 3 tons per hectare.

The Draft Program specified a wide range of agrotechnical steps to achieve these increases: more fertilizer, better irrigation, tool reform. The program also outlined a variety of extension programs, including education, demonstration, and use of models.

The Role of Mechanization in Intensification

The intensification strategy recognized that mechanization, based on tractors and other modern machinery, could not possibly be accomplished in the near future. Careful, convincing surveys by economists in 1956–57 documented this. However, Chinese economists discovered that intermediate technology ("semimechanization") could be crucial in breaking seasonal labor constraints. Some of the problems with rapid tractorization were specified by Chao Hsüeh, writing in *Planned Economy* in 1957.[5] He explained that to turn out the estimated need of 400,000 tractors would take at least 27 years, given the plans for industrial production. The tractors and implements would require about 5 million tons of iron and steel, a very substantial share of China's production. China's projected tractor fleet would consume about 4 million tons of petroleum products a year, and it was unknown when China would be able to produce that much. Leaving aside the question of the ability of the industrial sector to supply equipment for mechanization, the agricultural sector could not absorb machinery very rapidly. The farm land was divided into small plots which were inconvenient for mechanization. To the extent that the small plots reflected the system of private ownership, consolidation was possible. Graves and trees could be moved gradually. But to the extent that small plots reflected the need to terrace hillsides to control precisely the depth of water in rice paddies, change would be difficult. Mechanization was further complicated by the complex pattern of intercropping practiced in some regions.

Chao pointed out that there remained irrational, "blind" pressure for mechanization. The source of this pressure appeared to be provincial political leaders who, according to Chao, did not understand agriculture and farm mechanization. They had not made detailed surveys or experiments with regard to local conditions, labor power, or animal power. Some political leaders felt that only agricultural mechanization could guarantee that farm cooperation would be profitable. Presumably, the provincial leaders to whom Chao referred were expecting the central government to supply agricultural machinery at a subsidized price.

Quite apart from the *feasibility* of agricultural mechanization was the question of its *desirability*. Some Chinese economists came to the conclusion that even with plans to intensify agriculture, farm mechanization was irrelevant because China already had enough labor supply; machinery would only replace labor and not raise yields. In 1968, Red Guards quoted from speeches made by Po I-po and Liao Lu-yen in 1956, implying that mechanization was inappropriate for China for this reason. Po I-po said: "With such a large reservoir of manpower in the Chinese countryside and such complicated farming systems, it is impossible to in-

5. Chao Hsüeh, "The Problem of Agricultural Mechanization in China," *Chi-hua Ching-chi*, no. 4 (April 9), 1957 (ECMM 87, p. 10).

troduce mechanization. . . . If mechanization is introduced, the problem of surplus labor power in the countryside will become so acute as to defy solution.''[6] Liao Lu-yen, then Minister of Agriculture, said: ''With the exception of those areas where land is plentiful and labor power is inadequate and also with the exception of a number of economic crop growing areas . . . any simple and hasty steps to achieve mechanization are unacceptable to the masses, nor will they be conducive to raising agricultural output.''[7] According to Red Guard sources, ''a small handful of capitalist roaders'' submitted a report to the Party Center and to Chairman Mao on March 12, 1957. The report said that ''the development of farm mechanization is often the consequence of manpower shortage. The important role of farm mechanization is to raise labor productivity, but its effect is insignificant where the raising of the per unit area output is concerned. . . . Therefore, in view of the rich manpower and material resources of our country, the technical policy of farm mechanization becomes debatable.''[8]

Red Guards also said that ''preposterous views opposing farm mechanizations arose everywhere,'' and cited these arguments against mechanization: (1) Population is large, land is scarce, and intensive farming is practiced; (2) mountains and rivers impede machinery; (3) iron, steel, and petroleum are in short supply; (4) mechanization would not raise yield or total output; (5) mechanization would be too expensive.[9]

This argument was challenged by a sophisticated report authored by Huang Ching, who argued that selective mechanization and intermediate mechanical technology could be very important in increasing yields.[10] The report reflected an investigation by the State Technological Commission of which Huang was chairman, and was remarkably frank about the ignorance concerning agricultural mechanization at the highest levels of government:

Recently we sent a group of cadres and technicians to investigate the conditions in rural areas, and further exchange views with the ministries of Agriculture, Forestry and Water Conservancy respectively. Thus we began to know some of the

6. ''History of Struggle,'' SCMM 633, p. 10. 7. *Ibid.*. 8. *Ibid.*

9. *Ibid.*

10. Huang was also minister of the First Ministry of Machine Building and had been mayor of the technologically advanced city of Tientsin. He had been an important communist student leader in the 1930's. His report was published in two places: ''The Problem of Farm Mechanization in China,'' *Jen-min Jih-pao,* Oct. 24–25, 1957 (SCMP 1662, p. 13), and ''On Agricultural Mechanization in China,'' *Chi-hsieh Kung-yeh,* Nov. 6, 1957 (ECMM 120, p. 34).

concrete conditions in the rural areas which differed considerably from what we had imagined in Peking.[11]

[Before the investigation] we only knew superficially of some general conditions relating to farm mechanization in the Soviet Union and the United States, and considered in a general way that mechanization would raise labor productivity and agricultural production. . . . We did not seriously study the concrete conditions in the Chinese countryside, but only made deductions on the basis of agricultural mechanization in foreign countries.[12]

Huang agreed completely with the argument that China had to focus attention on increasing yields: "China, though expansive in area, is short of tilled land and reclaimable wasteland. Since she has a teeming population she has to adopt complex methods to increase agricultural production through intensive tillage and fine cultivation." Huang observed, however, that mechanization did not necessarily displace labor. He pointed out that even densely populated regions frequently suffered acute labor shortages in busy seasons, especially where multiple cropping was practiced. Labor supply was actually a constraint, and this problem was becoming more serious as multiple cropping was encouraged in regions with shorter growing seasons. As cropping systems were changed, the requirements for irrigation and fertilization also increased, creating a great need for machines and implements that could be used for transportation, water lifting, harvesting, threshing, plowing, transplanting—the operations that had to be done in the process of completing one crop and commencing the second. If machines were used carefully and selectively, according to Huang, they would not compete with human labor.

Moreover, a planned expansion of sideline industries would absorb the labor set free by the machines at other seasons. Huang described one successful cooperative that used machines to free labor to work at vegetable-oil processing, sugar refining, wheat-flour processing, brick making, carpentry, iron work, tailoring, animal raising, embroidery, horse-cart making, building, chicken raising, hay storing, and operating a grinding-stone house. Over 15 per cent of the peasants were engaged primarily in these sorts of enterprises.

The finding that mechanization permitted intensification of cultivation and differentiation of the economy agreed with what Mao had predicted. In *Socialist Upsurge in China's Countryside,* Mao had written:

11. Huang, "The Problem of Farm Mechanization," SCMP 1662, p. 14.
12. *Ibid.*, p. 13.

After mechanization, even a larger amount of labor power will be saved. Will there be no outlet for labor power then? There will still be an outlet, according to the experiences of some mechanized farms. The reason is that when the scope of production has expanded, production departments are increased and work is further carried to greater detail; and so there is no fear that labor will find no use.[13]

In addition to the intensification of agriculture implied in this statement, Mao envisaged diversification of tasks leading to new jobs:

After the mechanization of agriculture . . . there will emerge in the future various kinds of undertakings never before imagined by people, and the yields of agricultural crops will be raised several times and even scores of times the present level. The development of industrial, communications and exchange enterprises will even be beyond the imagination of the people of the past. Likewise, there will be such developments in the fields of science, culture, education and public health.[14]

To meet China's needs, machinery had to be designed afresh, Huang's report discovered. Specifically, it urged these features be taken into account: (1) The tremendous diversity in China's physical geography, and cultivation practices meant that a great variety of implements and machines were required. (2) Machines should be sturdy, simple, and easy to repair. (3) Efforts should be made to use old-fashioned implements with modern machines rather than abandoning them. (4) Whenever possible, machines should have multiple uses. They should be able to work more than one crop, and they should be able to run pumps and processing equipment as well as cultivate the fields. (5) The price of machinery should be kept low. (6) Efforts should be made to use fuels other than petroleum. Moreover, the provision of spare parts and repair service was essential. The report quoted a cadre in a cooperative who observed, "When the peasants do not work, we preach Marxism-Leninism to mobilize them. When the draft animals do not work, we whip them. But when the machines do not work, we are at our wits' end."[15]

One result of this re-evaluation was that the expansion of tractorization slowed (see Table 5.1). The growth of the tractor supply in 1957 (27 per

13. Comment on article "Outlets Found for Surplus Labor Power," cited by P'ang Hsien-chih, *Upsurge of Socialism in China's Countryside: Document of Historical Significance in the Party's Leadership of the Socialist Revolution in the Rural Areas* (Peking: Hsüeh-hsi, 1956), CB 388, p. 27.

14. Comment on article "Mobilize Women to Take Part in Production to Solve the Difficulties of the Insufficiency of Labor Power," cited in *ibid.*, p. 28. Also available in a slightly edited version in *Socialist Upsurge in China's Countryside* (Peking: Foreign Languages Press, 1957), p. 286.

15. Huang, "The Problem of Farm Mechanization," SCMP 1662, p. 23.

Table 5.1. Farm tractors in China, 1952–60

Year	(Standard) tractors in use	% increase
1952	2,006	
1953	2,719	35
1954	5,061	86
1955	8,094	60
1956	19,367	140
1957	24,629	27
1958	45,330	86
1959	59,000	30
1960	79,000	34

Source: Chao, *Agricultural Production,* p. 107.

cent) was the smallest in the decade. (Because all tractors were imported before 1958, the growth rates immediately reflected policy shifts.)

Mao Sums Up Mechanization Policy

Importing fewer tractors, however, did not mean lack of interest in mechanization. At this time, the Chinese leadership was showing increased interest in the broad expansion of agricultural technology. Mao envisaged a technical revolution following in uninterrupted fashion the socialist revolution that had just been completed in industry and agriculture. The mood of the national leadership, meeting at Hangchow and Nanning in January 1958, was summarized by Mao. He called for "universal popularization of experimental plots," and urged, "We must learn new skills, understand our functions in the real sense and know science and technology." Regarding farm mechanization, Mao offered specific suggestions concerning farm tools and stressed research and development, with an emphasis on medium-sized and small tools.

> Farm implements research centers should be established in provinces, autonomous regions and municipalities directly under the Central Government. They should assume special responsibility in conducting research in improved farm implements and medium and small mechanized farm implements. They should establish close ties with farm implements plants and hand over their research results to the latter for manufacture.[16]

Another aspect of this strategy was developed around this time, namely that farm-tool reform and agricultural mechanization would be closely as-

16. Mao Tse-tung, "Sixty Work Methods [Draft]," Jan. 31, 1958, article 53; available in *Long Live Mao Tse-tung Thought* (n.p., n.d.), CB 892, p. 13.

sociated with a policy of emphasizing the expansion of rural local industry. While a detailed analysis of policies toward industrialization is outside the scope of this study, the issue must be touched upon because it overlaps significantly with agricultural mechanization policy. During the First Five-Year Plan (1953–57), the focus of industrial development strategy was the creation of a few centers of modern heavy industry, utilizing advanced technology. Once this base was established, would future expansion of industry be located in the same centers? While there might be some internal efficiencies, some Chinese leaders undoubtedly feared that such a strategy would result in excessive urban development, producing a situation in which there was inadequate sharing of the benefits of industrial productivity with the rural sector. At a conference of Party leaders held in Chengtu in March 1958, Mao proposed linking the questions of farm mechanization and industrial development by encouraging rural local industry to manufacture improved tools and farm machines.[17] This is specified in Paragraph D of the document quoted below.

The overall policy toward agricultural mechanization was summarized at the Chengtu Conference in a report entitled "Views on the Question of Agricultural Mechanization," and stands as a watershed in China's agricultural mechanization policy. The full document is not available, but substantial fragments have been quoted by Red Guard groups, so that its thrust is known. In the following reconstruction, the order of the fragments and the headings have been supplied by me, but the content comes from Red Guard quotations. After some sections, there is, enclosed in parentheses, a Red Guard commentary on the preceding sentences.

Views on the Question of Agricultural Mechanization

A. Importance of Agricultural Mechanization

The meeting is in complete agreement with Chairman Mao's directive concerning the movement for the reform of farm implements. The mass movement for the reform of farm implements with the broad peasantry participating in it is the germination of technical revolution and a great revolutionary movement. All places throughout the country should universally and actively popularize it, and bring

17. A Soviet analyist suggested that China emphasized medium- and small-sized enterprises because of the country's extreme backwardness in technology and economy, a large population, and an employment problem. Mao rejected this argument, and said that small- and medium-scale industry could be justified on productivity alone: "We are developing a large number of medium and small-sized enterprises under the guide of big enterprises and adopting extensively indigenous methods under the guide of foreign methods mainly for the sake of achieving a high rate of industrialization" (Reading Notes on the Soviet Union's *Political Economics, Miscellany*, p. 261).

about gradual transition to semi-mechanization and mechanization through this movement.[18]

B. Relationship between Mechanized, Semi-mechanized, and Improved Tools

Reform of farm implements, regardless of whether they are mechanized (including power-operated machines and machine-drawn farm implements), semi-mechanized (so-called new-type animal-drawn farm implements) or preliminary improved from old-type farm implements (so-called improved farm implements), is in all cases conducive to the raising of productivity of agricultural labor. We should not simply wait for agricultural machines and relax popularization of new-type animal-drawn farm implements and improved farm implements.

Main emphasis should be given to small-sized farm machines, to be coordinated with large and medium-sized machines of appropriate quantity.[19]

C. Local Financing and Management

Agricultural cooperatives must principally rely upon their own strength to realize agricultural mechanization. Only in this way can the undertaking of agricultural mechanization be run to achieve greater, faster, better and more economical results.[20]

(In this way it is possible not only to exploit the enthusiasm of the collective economy, but also to basically solve the contradiction that the peasants have no farm machines—to unify the ownership of agricultural machines and the right to use these machines, to closely link agricultural mechanization with the demand of agriculture, to draw unified plans for agricultural production, and to fully use and properly manage the agricultural machines.

Meanwhile, it facilitates the transition from basic ownership of a production team to the basic ownership of a commune, transition from communal ownership to state ownership, and consolidation and development of people's communes.

Practice has fully proved that the collective economy of the people's communes has a very high enthusiasm and very great potentials in undertaking agricultural mechanization, and that many people's communes can, within a relatively short period of time, raise funds, build their workshops, train their operators by their own effort, and carry out agricultural mechanization on a relatively big scale.

The people's commune not only can afford to buy agricultural machines, but also can use and manage them well.)[21]

18. Committee Taking over the Control of the Department of Agricultural Mechanization of Heilunkiang, "Let the Radiance of Mao Tse-tung's Thought for ever Shine over the Road of Agricultural Mechanization," *Nung-yeh Chi-hsieh Chi-shu,* no. 4 (July 8), 1967 (SCMM 600, p. 5).

19. Peking Agricultural Machinery College, East is Red Commune, Criticism and Repudiation Office, "Completely Settle the Heinous Crimes of China's Khrushchev and Company in Undermining Agricultural Mechanization," *Nung-yeh Chi-hsieh Chi-shu,* no. 5 (Aug. 8), 1967 (SCMM 610, p. 24). Also available in *ibid.,* SCMM 600.

20. *Ibid.,* SCMM 600, p. 3.

21. Eighth Ministry of Machine Building, Revolutionary Great Alliance Headquarters and Revolutionary Great Criticism and Repudiation Group of Organizations, "Two Diametrically Opposite Lines in Agricultural Mechanization," *Nung-yeh Chi-hsieh Chi-shu,* no. 9, 1968 (SCMM 633, p. 41).

D. Local Production of Tools

In regard to manufacture of farm machines (including machine-drawn farm implements, new-type animal-drawn farm implements and improved farm tools), with the exception of large farm machines and those with relatively high technical standards, generally the emphasis should be given to local industry. Arrangements may be made by the various localities or through the various coordinated working zones in the light of local conditions and needs.[22]

(In this way, first, it facilitates the exploitation of enthusiasm locally and provides favorable conditions for making farm tools by adopting means appropriate to local conditions and for serving agriculture in a better way.

Secondly, it proceeds from the great strategic idea "Prepare for war and natural calamities and prepare everything for the people."

Thirdly, the local industry comes close to the countryside, so that the situation is favorable for consolidating the worker-peasant alliance and gradually eliminating the difference between workers and peasants.)[23]

E. Role of Central Government

The relevant departments of the central government should help local industries with respect to techniques and exchange of experience. In those places where the local industry is relatively weak in foundation, and where this problem cannot be solved by mutual readjustments between provinces in a coordinated working zone, the relevant departments of the central government should work out the readjustments.[24]

Farm tool reform should be carried out in a manner consistent with local characteristics, and not in a uniform manner.[25]

Thus by early 1958 a strategy had emerged for developing China's agriculture. The strategy stressed intensification of cultivation, achieved partially through increased labor inputs but also through selective mechanization and tool reform. It envisaged the manufacture of improved farm tools and farm machinery in local factories, which would be developed to meet this need. Agriculture would, of course, remain organized according to principles of collective ownership, and the improvement of farm tools would strengthen the collective economy.

22. "Completely Settle the Heinous Crimes," SCMM 610, p. 27.
23. "Two Diametrically Opposite Lines," SCMM 633, p. 41.
24. "Completely Settle the Heinous Crimes," SCMM 610, p. 27.
25. "History of Struggle," SCMM 633, p. 25.

6. Agricultural Mechanization Policy in the Great Leap

The Theory of the Great Leap

How would this vision of agricultural change—stressing intensification and gradual mechanization combined with collective ownership of machinery—be implemented? A remarkable method was adopted during the "Great Leap forward" of 1958–60 which included several dimensions. Most important was the idea that massive increases in *labor inputs* would create large increases in production and generate a surplus which would make increased investment possible. Second, rapid and radical *social reorganization*—including the formation of people's communes—was designed to permit increases in labor inputs. Third, *mass participation* in a tool-reform campaign was to be utilized to break through the labor constraints. It was hoped that by one massive effort the country could be propelled in just a couple of years into modernity, with an advanced economy and with personal relations and incentives based on the communist principle of "to each according to his needs."

This style of implementation did not work. Modernization cannot be achieved overnight. The problems were numerous.

First, the Great Leap strategy was built on an ambiguous theory of labor supply. On the one hand, it was based on the overall vision that the one resource in plentiful supply was human labor—the energies of roughly 550–600 million Chinese people. Some empirical studies implied (contrary to expectations) that labor had an increasing marginal utility.[1]

1. In one commune, the following data were reported to support the contention that intensified cultivation not only increased yield, but increased yield at a greater rate than the increased labor:

	man-days/ha.	*yield kg./ha.*	*kg./man-day*
Regular cultivation	.71	105	6.5
High-yield cultivation	3.54	1,300	16.3

Although there were several reports of this type, it is possible that the Communist leadership did not analyze them with the necessary skepticism. The surveys are reported by Liu Hsi-keng, "A Survey of the 'Sow Less and Yield More' Program," *Hung-ch'i,* no. 13 (Dec. 1), 1958 (ECMM 157, p. 36).

Labor was considered virtually a free good, which could be increased almost indefinitely. At precisely the same time, however, many Chinese economists were arguing that there was a labor *shortage*. They predicted that intensive cultivation would require doubling or tripling of labor inputs (plowing, transporting seed and fertilizer, sowing, transplanting, etc.). There was general agreement that the labor supply was simply not adequate to double or triple the labor input. Labor shortages were reported all over China as intensive cultivation was tried.[2]

The most widely discussed solution to the sudden labor shortage was social reorganization. Several agricultural cooperatives were merged with local government organizations to create a new rural institution, the people's commune. It was thought that by pooling labor into larger groups, by organizing labor in military fashion so that brigades could be assigned flexibly throughout a commune, and by collectivizing certain household tasks such as cooking, sewing, and childcare, labor power could be released from various activities and utilized for farming. Communes were also attractive from an ideological point of view; there was less private ownership of property and more collectivization of private and family life. There may have been foreign policy considerations as well. By establishing communes China was advancing to communism more rapidly than the Soviet Union and thereby implicitly challenging the Soviet Union's leading role.[3]

The idea that labor inputs could be increased substantially through

2. One estimate was that China as a whole had 33 billion work days available, but that this was only about half of the labor required. Feng Chih-kuo, "How to Solve the Problem of Labor Shortage on the Agricultural Front," *Ching-chi Yen-chiu,* no. 3 (March 17), 1959 (ECMM 167, p. 18). A report on labor shortages in Hunan appeared in Yang Ta-chih, "Some Questions of Technological Revolution in Agriculture," *Hsüeh-hsi,* no. 9 (May 3), 1958 (ECMM 134, p. 33). Labor shortages in Honan were reported by a survey by the Agricultural Section of the Economic Investigation Corps of the Third-Year Students of the Economic Department, Wuhan University, "The Road to the Realization of the Mechanization of Agriculture—Report on the Investigation of the Chao-Ying People's Commune, Shangcheng, Honah," *Li-lun Chan-hsien,* Feb. 10, 1959 (ECMM 167, p. 25). This perspective on the rural economy is more consistent with recent Western theories, including those of Theodore Schultz, suggesting that the traditional agricultural economy uses its resources (including labor) quite efficiently, considering the technologies that are available.

3. Mao was well aware of foreign policy implications, and after the Quemoy crisis of August–September 1958, in which the United States threatened to use nuclear weapons against China and the Soviet Union refused to include China under its "nuclear umbrella," Mao observed: "It would not look right for China to enter communism ahead of the Soviet Union. . . If we rush, we may possibly commit errors in international matters" (Mao's Speech at the Sixth Plenum of the Eighth Central Committee, Dec. 19, 1958, *Miscellany,* p. 145).

social reorganization, while true in theory and eventually also in practice, was exceedingly difficult to implement. Greatly increasing labor inputs required an exceedingly effective incentive system, but the incentive system adopted at the time was not suitable.[4] Collective dining rooms were established in many villages and people could eat as much as they wanted. Likewise clothing and medical and other services were given freely, according to need. Very few preasants were willing to accept such a rapid transition to the communist principle of distribution according to need, with little ideological preparation and with little assurance of how everyone else in the village would respond. Without labor incentives that fit the expectations of the peasants, labor enthusiasm suffered; the peasants in many cases virtually went on sit-down strikes. Moreover, tremendous problems were encountered when boundaries of social organization were shifted so that natural marketing and economic units were no longer congruent with administrative and production units. People who previously had not worked together were suddenly expected to cooperate.[5]

From a technological point of view, the belief that economic problems could be solved through massive increases in labor inputs proved to be wrong. Intensification of agriculture required chemical fertilizer, insecticides, pumps, new sources of energy, and new agricultural machines (to save time between crops). Increased labor proved complementary to these inputs; in their absence, increased labor often had negative effects, and these modern inputs could not be supplied overnight.

Additional problems may have been caused by the premature popularization of newly developed high-yielding varieties of rice, which were not drought-resistant and suffered grievously in 1959–60 when bad weather and managerial difficulties combined to disrupt water supplies.[6]

Administration of the Great Leap also presented problems. To achieve overnight modernization, administrators often applied uniform policies to regions with great ecological and social differences; or they selected a

4. Mao's views on an appropriate incentive system are quite complex. He is frequently described, correctly, as putting great emphasis on moral incentives. However, Mao balanced moral and material, personal and collective incentives. He pointed out about the Soviet Union: "In the Stalin period, they placed far too much emphasis on collective interests and paid little or no attention to personal income, placed far too much emphasis on public interests and paid little or no attention to private interests. Today they have gone to opposites. They overemphasize material [personal?] interests and pay little attention to collective interests" (Reading Notes on the Soviet Union's *Political Economics, Miscellany,* p. 290).

5. G. William Skinner, "Marketing and Social Structure in Rural China," *Journal of Asian Studies* 24 (1964–65), 1–43, 195–228, 363–399.

6. This observation has been made by H. V. Henle in correspondence to me.

few specific techniques for popularization, and then gauged their success simply by the rapidity by which they were popularized. Such a simplistic administrative approach could not work; only disruption, imbalances, and mistakes could result.

With all these problems, the Great Leap made two immeasurable contributions. First, it marked a clear recognition that that agriculture would have to be developed. The overall economic strategy of the First Five-Year Plan to stress heavy industry had created serious intersectoral imbalances, and some new strategy was needed.[7]

From a psychological point of view, the Great Leap made clear that changes could be made. It projected the vision of agricultural transformation, industrial development (in both urban and rural areas), and social advance. Even though some projects begun during the Great Leap were later suspended because of lack of resources or personnel, the idea of change had been implanted. A few years later, when the economy stabilized and more resources were available, many of the original projects were resurrected and carried out under sound management. Numerous schools, factories, and innovations in industry and agriculture which bore fruit in the middle and late 1960's were originally started this way. If a major problem in development is projecting a vision of change, then the Great Leap, with all its problems, made a crucial contribution.

Mao Tse-tung's own summary evaluation of the Great Leap, stressing the backyard blast furnaces, is very similar to that of Western observers:

> If one looks at [the Great Leap Forward] from the viewpoint of the law of value alone, the conclusion that he will inevitably draw is that we lost more than we gained. He will inevitably describe last year's all-out campaign for smelting steel and iron as ineffectual labor, low-quality steel produced by indigenous methods, big country and many subsidies, indifferent economic results, etc. From the partial, short-term viewpoint, the all-out campaign for smelting steel and iron seems like it has cost us dear.
>
> However, viewing the situation as a whole in the long term, it is worthwhile because the all-out campaign for smelting steel and iron has opened up a new phase in the overall economic construction of our country. The establishment of a large number of new steel and iron bases and other industrial points throughout the country will make it possible for us to go a long way to speed up our tempo.[8]

7. Victor Lippit, "The Great Leap Forward Reconsidered," *Modern China* 1:1 (Jan. 1975), 92–115.

8. Mao Tse-tung, Reading Notes on the Soviet Union's *Political Economics, Miscellany*, p. 285.

The Mass Tool-Reform Campaign of 1958

Recognizing that a labor shortage did exist (in comparison to the stupendous targets put forth) and that social reorganization would not solve the labor problem, the leadership also inspired a remarkable tool-reform campaign, marked by extensive mass participation in developing and diffusing intermediate technology. The potential of new tools had been reportedly displayed during the irrigation campaign of the previous year, when new tools had been invented to reduce the manpower needed to move earth.[9]

The overall strategy for the new mechanization campaign was determined at the Chengtu Conference in March 1958. The conference placed priority on intermediate technology (gradual mechanization and popularization of animal-drawn equipment) manufactured locally. During the Cultural Revolution, Red Guards claimed that this decision was not publicized.[10] Whether or not the text of the Chengtu decision was released, its general ideas were popularized immediately. On March 15, 1958, a directive was issued jointly by the First Ministry of Commerce, the Ministry of Agriculture, the All-China Federation of Handicraft Producers' Cooperatives, and the General Office of the People's Bank. The directive encouraged cooperatives to use new tools, asked that new machines be designed, produced, and funded locally, and encouraged development of alternative energy sources. The directive did not, however, unleash a mass campaign. Rather than speaking of liberating the masses' creative powers, the directive had a tone of discipline and reminded peasants to keep their tools in good repair.[11]

9. "Mass Movement to Improve Farm Tools," NCNA Peking, April 27, 1958 (SCMP 1763, p. 33).

10. Red Guards charged: "China's Khrushchev and his jackals imposed a strict blockade on this important document and refused to transmit it downward for as long as seven years. It was not until July 1964 when some people discovered this document that the followers of the bourgeois reactionary line in the Eighth Ministry of Machine Building [which supplied agricultural machinery] had to pass it around among cadres holding the post of division and bureau directors and higher ranks but they still refused to transmit it to the broad masses" (Peking Agricultural Machinery College, East Is Red Commune, Criticism and Repudiation Office, "Completely Settle the Heinous Crimes of China's Khrushchev and Company in Undermining Agricultural Mechanization," *Nung-yeh Chi-hsieh Chi-shu,* no. 5 [Aug. 8], 1967; SCMM 610, p. 19). This charge contains one minor factual error. The Eighth Ministry of Machine Building had not been created by July 1964; it was created in January 1965. Presumably, the Red Guards were referring to the Ministry of Agricultural Mechanization.

11. "Joint Directive Issued on Farm Tools," NCNA Peking, March 15, 1958 (SCMP 1758, p. 11).

The Party Central Committee and the State Council issued another directive on July 14 asking that tools receive increased attention. Emphasis was placed on the need to popularize new tools.

> The problem at present is that although there have been many inventions, not enough has been done to promote their use and the implements now in extensive use in rural areas are still those of the old and low-efficient types.
>
> As a result, the contradictions between the effort for bumper harvest and the shortage of labor has become increasingly acute with labor stringency preventing, in some areas, a full realization of production measures. Under the circumstances, rapid promotion of agricultural implements has come to be one of the key measures to ensure all round bumper agricultural harvest.[12]

Perhaps this directive was accompanied by instructions to take it seriously and treat it as a major campaign. Within a few days, tool-innovation committees were set up by Communist Party committees in the rural areas (in at least some localities) to lead the campaign by criticizing ideas of "purely expecting state supply of machinery" and "asking for money, materials and machinery from the state." They were also supposed to provide an impetus for making new tools by organizing the carpenters and other workmen and commandeering materials to develop tools.[13] The importance of leadership by the Communist Party in this effort was stressed in one report:

> Tool reform in this district has invariably been carried out under the immediate command of the first secretaries of the Party committees at different levels, and has been regarded as the most important of all productive measures. Technical revolution committees have been set up in all parts of the district, and competent cadres have been assigned to supervise tool reform, and to see that it is carried out on a large scale and with perseverance.[14]

Any doubts concerning the priority the center attached to tool reform must have been dissolved when P'eng Chen and T'an Chen-lin, acting on Mao's direct instructions, telephoned provincial Party committees on August 20 and asked for progress in the tool-reform campaign.[15] Particu-

12. "CCPCC and State Council Issue Directive on Movement for Improvement of Agricultural Implements in Rural Areas," NCNA Peking, July 14, 1958 (SCMP 1819, p. 5).
13. "Buds of Technical Revolution in Agriculture Shoot Forth Everywhere," NCNA Peking, March 20, 1958 (SCMP 1744, p. 5).
14. "Fuyang Administrative District Sums up Experience," *Jen-min Jih-pao,* Jan. 21, 1959 (SCMP 1951, p. 23).
15. "Telephone Conferences Held on Tools Improvements; Instructions Given by Chairman Mao," *Jen-min Jih-pao,* Aug. 21, 1958 (SCMP 1845, p. 20).

lar importance was attached to the installation of ball bearings in transportation vehicles, processing machines, and waterwheels. At the same time, a *People's Daily* editorial cited another joint directive from the Central Committee and State Council stressing in even stronger language the importance of tool reform: "The most important and urgent problem arising from the basic completion of socialist revolution in our country consists in rapid development and elevation of the productive force of the society, and *the central link* in the chain of developing the productive force of agriculture in our country is to substitute improved, new farm tools for old and unwieldy farm tools."[16] The immediate results of this central pressure were mixed. There was instant statistical success with regard to a few specific tools. Within three weeks there were reports that 1,530 counties (70 per cent of the total) had set up 300,000 ball bearing workshops, which had made 41.5 million sets of ball bearings. By January 1959, 122 million sets of ball bearings were reportedly produced.[17] In July and August 1958, it was claimed, 154 million improved farm tools went into use—more than one tool for each family.[18] (During this period the statistical reporting system fell apart and many statistics reflected gross exaggerations. There is no way of knowing the accuracy of these or other statistics issued during this time, and they must be used with extreme caution.

In fall 1958, a new tool, the cable-drawn plow, was developed to permit deep plowing in small plots and paddy areas, where the double-blade double-wheel plow was unsuitable. The cable-drawn plow utilized any energy source (including human, animal, water, wind, internal combustion engine, electric motor) to turn a winch which could pull a cable attached directly to a plow. It appeared to offer a method of applying mechanical energy to paddy cultivation without the weight, expense, and inefficiencies of tractors. T'an Chen-lin, politbureau agricultural expert,

16. "A Great Campaign for Improving Tools," *Jen-min Jih-pao* editorial, Aug. 21, 1958 (SCMP 1845, p. 21). Emphasis added.

17. "1958 Tools Innovation Movement," NCNA Peking, Jan. 2, 1959 (SCMP 1929, p. 32). An enthusiastic report and description of the campaign to install ball bearings in carts, wheelbarrows, waterwheels, etc. is offered by Anna Louise Strong: Metal bearings were hammered out by hand; some ball bearings were made of glass, porcelain, and even acorns. For the acorn bearings, the races were made of bamboo. (*The Rise of the People's Communes in China* [New York: Marzani and Munsell, 1960], p. 42).

18. "Progress of China's Campaign to Improve Tools," NCNA Peking, Sept. 10, 1958 (SCMP 1858, p. 15).

attended a conference on the cable-drawn plow in September 1958 and decided that it should be immediately popularized. Reports said that 150,000 were made by mid-September and 600,000 more would be ready soon thereafter.[19] By early November 2.4 million such machines were said to have been constructed.[20] By January 1959 the statistical success was greater: cable-towed machinery was up to 2.6 million.[21] Unfortunately, the cable-drawn plow was unsuccessful, probably because of insufficient energy sources and because the cables wore out quickly. Red Guards later criticized this impetuous popularization:

> In September 1958, in respect to the tentative model of a cable-drawn tractor called "chiaokuan" which had not been tested, [T'an Chen-lin] said: "These deep-tillage plows should be universally popularized throughout the country within one month . . . [the task] must be completed within a time limit. . . . If [the task is] not completed, we will look for you!" Moreover, he directed that progress in popularization be reported in *People's Daily* once every five days. He ordered arbitrarily that these "chiaokuan" machines be used to till the land. As a result, the farm tool reform mass movement was grossly impeded.

This premature popularization of the cable plow was termed "ultra-left" and "actually extremely right in essence."[22]

In the case of ball bearings, no specific criticism of the campaign have been found. On the other hand, it was never referred to later as an example of success. It seems to have been forgotten.

As a result of the tool-innovation campaign, it was estimated that 50 billion man days were saved throughout 1958. This would indicate that farm work accomplished more than doubled, because estimates of the total work days available for agricultural labor range from 33 to 47 billion days.[23] Some regions had set up many small, primitive machine shops.

19. "New 'Towing Cable' Machinery Recommended for Agriculture," NCNA Nanking, Sept. 19, 1958 (SCMP 1867, p. 26).

20. "Results of Farm Tools Innovation Campaign," NCNA Peking (based on the day's *Jen-min Jih-pao*), Nov. 4, 1958 (SCMP 1891, p. 12).

21. "1958 Tools Innovation Movement," NCNA Peking, Jan. 2, 1959 (SCMP 1929, p. 32).

22. "History of Struggle," SCMM 633, p. 16.

23. "Conference Sums up Movement to Improve Farm Tools," NCNA Peking, Dec. 22, 1958 (SCMP 1922, p. 9). Peter Schran has estimated that to total labor pool for 1958 was 47.5 million labor days (*The Development of Chinese Agriculture, 1950–1959* [Urbana: University of Illinois Press, 1969], p. 75). A Chinese estimate was that roughly 33 billion work days were available (Feng, "How to Solve the Problem of Labor Shortage," ECMM 167, p. 18).

More inanimate energy was being used, in the form of waterwheels, windmills, and marsh gas.[24]

One of the problems encountered with this campaign was the fact that not everyone supported it. Some peasants and cadres demanded full mechanization with tractors and complete sets of implements; they saw no value in creating improved tools which would soon be obsolete. The July 14 directive had warned that "ideas which suggest 'postponement until full mechanization is realized' or 'until the implements are improved to full perfection' are wrong."[25] Another report also noted this tendency:

> Certain people, when asked to increase the productivity of labor, would ask the government for all kinds of agricultural machinery. These people know very well that the government cannot be expected to meet their demands for machines at once, but they would rather wait passively for supplies than carry out innovations and improvement on existing tools by mobilizing the activism and creativeness of the masses.[26]

It is interesting that, both at this juncture and in 1954, opposition was expressed very indirectly, by asking for impossible supplies of advanced machinery. Presumably these people did not accept the basic premise of the tool-reform campaign, namely that labor still had a positive marginal utility and could raise production and the standard of living.

Another problem, implied above, was the tendency of the central leadership to latch onto particular innovations (such as the cable plow or ball bearings) and expect them to be popularized universally. The particular innovations may or may not have been sensible (in this case it appears that they were not) but even the best innovation cannot be implemented everywhere. In this case, as in others, the erroneous popularization of inappropriate tools probably weakened the program significantly, not only by wasting materials, but more importantly by unnecessarily associating the legitimacy of the leadership and the theory of farm tool reform with a mistake.

In actuality, few of the newly invented tools (other than those pushed by the central leaders) were mass-produced and successfully integrated

24. "Conference Sums up Movement to Improve Farm Tools," SCMP 1922, p. 9; Li Ch'ing-yü, "A Summary of the Conference on Agricultural Mechanization and Electrification," *Nung-yeh Chi-hsieh*, no. 1 (Jan. 15), 1959 (ECMM 161, p. 12).

25. "CCPCC and State Council Issue Directive," SCMP 1819, p. 5.

26. Hsieh Yin-ch'i, "Ways to Increase Productivity of Labor in Agriculture in our Country," *Jen-min Jih-pao*, March 12, 1959 (SCMP 1986, p. 2).

into production.[27] On the average, only 800 tools of each new design were made and distributed.[28] One reason is that many were poorly manufactured. Li Ch'ing-yü, an assistant to Minister of Agriculture Liao Lu-yen and director of the Chinese Academy of Agricultural Sciences' Agricultural Mechanization Institute, explained: "Many farm tools produced were of poor quality, breakable and not durable. Some localities fitted tools with ball-bearings in three days, and gave them up after five days. The reasons are that time was too short, tasks were too heavy, facilities were poor and subjectively quality was not given due attention. This is indeed a drawback."[29]

Despite these problems, Li remained broadly enthusiastic about tool reform and the mass movement: "The tools innovation campaign has convinced us that we can only rely on the masses for mechanizing and electrifying our agriculture and that we should start with innovation of farm tools and pass over from semi-mechanization to full mechanization and electrification."[30]

Management of Machinery

The establishment of people's communes in the summer of 1958 provided an obvious solution to the nagging question of ownership and management of agricultural machinery. The Chengtu Conference of March 1958 had determined that "agricultural cooperatives must principally rely upon their own strength to realize agricultural mechanization."[31]

On May 3, 1958, the State Council acted explicitly on the question: "To place agricultural machinery under the management of agricultural producer cooperatives will facilitate the development of agricultural production. Please follow and sum up the new conditions and new problems arising from the transfer of agricultural machinery to cooperatives."[32]

Decentralization of agricultural machinery had begun even before the formal State Council order. Toward the end of March, it was reported

27. "National Conference of Directors of Tool Renovation Officers Ends," NCNA Peking, Feb. 19, 1959 (SCMP 1964, p. 15).

28. Feng, "How to Solve the Problem of Labor Shortage," ECMM 167, p. 18.

29. Li, "Summary of the Conference," ECMM 161, p. 12. 30. *Ibid.*

31. Committee Taking over the Control of the Department of Agricultural Mechanization of Heilungkiang, "Let the Radiance of Mao Tse-tung's Thought for ever Shine over the Road of Agricultural Mechanization," *Nung-yeh Chi-hsieh Chi-shu*, no. 4 (July 8), 1967 (SCMM 600, p. 3).

32. "Summing-up Report on the National Conference of Tractor Station Masters (Excerpts)," *Chung-kuo Nung-pao*, no. 5 (Nov.), 1958 (ECMM 147, p. 16).

that agricultural cooperatives were planning to buy tractors.[33] At the same time, two cooperatives in Szechwan purchased two tractors from tractor stations.[34]

The merger of cooperatives into communes created a rural unit of organization with a larger size and financial base. By the summer and fall of 1958, it was not cooperatives but communes that were purchasing tractors.

One report from the experimental site of Paich'uan County, Heilungkiang, indicated that commune management was even better than cooperative management. Hsingyeh Cooperative, which had received tractors in the 1957 test, joined Hsingnung People's Commune, and the increased scale of operation was reported to be beneficial. By rearranging and merging plots of land, operating costs were reduced to as low as ¥0.33 to ¥0.19/mou. (Exactly how there figures were computed was not stated.) In addition, the greater financial resources of the commune made purchasing equipment more feasible.[35] From Anhwei it was reported that commune ownership boosted the utilization rate of machines by 16 per cent and lowered the costs by 13 per cent.[36]

By the end of 1958, 15,102 standard tractors had been purchased by 1,700 communes.[37] This represented 33 per cent of China's total stock of tractors, and about 8 per cent of the communes. Roughly 70 per cent of the tractors owned by the MTS's at the end of 1957 were sold to communes;[38] in addition, the communes received new tractors.[39]

33. "Northeast Plans Farming Mechanization," NCNA Shenyang, March 23, 1958 (SCMP 1741, p. 21).

34. The terms of the purchase were ¥38,000, to be paid in installments; the tractor station would remain in existence and maintain the tractors, as well as give technical guidance. "Szechwan Agricultural Cooperatives Buy Tractors," NCNA Chengtu, March 21, 1958 (SCMP 1741, p. 29).

35. Work Group of CCP Nunkiang District Committee of Heilungkiang, CCP Paichuan County Committee of Heilungkiang, and Heilungkiang Branch of the New China News Agency, "The Universal Establishment of People's Communes Accelerates Mechanization—An Investigation into Hsingnung People's Commune in Paichuan Hsien, Heilungkiang Province," *Jen-min Jih-pao,* Dec. 4, 1959 (SCMP 2160, p. 11).

36. Representatives of Anhwei Province attending the Work Conference on "Grasping Revolution and Stimulating Production" called by the Eighth Ministry of Machine Building, "China's Khrushchev and His Ilk Guilty of Heinous Crimes in Undermining Farm Mechanization in Anhwei," *Nung-yeh Chi-hsieh Chi-shu,* no. 9, 1968 (SCMM 644, p. 32).

37. Li, "Summary of the Conference," ECMM 161, p. 18.

38. *Ibid.*

39. Red Guard groups offered a somewhat different statistic: "By the end of 1958, tractors operated by more than 2,300 communes represented 70 percent of the total number of tractors in the nation's agricultural system" ("History of Struggle," SCMM 633, p. 14). This would imply a total of 31,600 standard tractors were owned by communes. It is possi-

Most likely the decentralization of 1958 affected only the administrative and financial aspects—that is, the commune assumed managerial and financial functions. The physical location of tractors probably did not change. In 1957, the average tractor station had 31 standard tractors. In 1958 and 1959 when communes bought tractors, they often bought in units of roughly that size. For example, a commune in Szechwan bought 46 tractors in the spring of 1959. A commune in Honan bought 19 tractors. One commune in the Northeast bought 20; another bought 38.[40] However, it is also clear that some of the tractors must have been detached from the stations, because 1,700 communes purchased tractors and there were only 383 stations in 1957. A commune in Kiangsu bought only 12 tractors, for example.[41]

The program of decentralization continued during 1959. By the end of the year, the communes of Heilungkiang owned almost 3,000 of the 10,363 standard tractors in the province.[42] Communes in Shensi established (from 1958 to 1961) over 100 tractor stations.[43] (No national-level statistics are available concerning the extent of commune ownership of tractors during 1959.) Throughout 1960 and 1961, there were continued sporadic reports that communes owned tractors, but no statistics are available.

During this period the MTS system continued to receive new tractors, so that despite selling about 8,400 standard tractors, the MTS system ended 1958 with roughly the same number of standard tractors that it had at the end of 1957—12,000–13,000.[44] It would seem likely that the new

---

ble that this latter figure is accurate, but more likely that the Red Guards misinterpreted to what the figure 70 per cent referred.

40. "Southwest China Commune Mechanizes Plowing and Sowing," NCNA Chengtu, April 13, 1959 (SCMP 1995, p. 11); "Southwest Hsien Advancing towards Mechanization of Agriculture," NCNA Chengtu, Nov. 11, 1959 (SCMP 2138, p. 27); "Central China Commune Paves Way for Mechanization," NCNA Chengchow, Sept. 18, 1959 (SCMP 2102, p. 35); "Northeast China Communes Buy Tractors," NCNA Shenyang, Nov. 4, 1959 (SCMP 2135, p. 26); "Tractors Boost Commune's Production," NCNA Harbin, Nov. 6, 1959 (SCMP 2136, p. 19); "Northeast China Province on the Road to Mechanization," NCNA Harbin, Nov. 8, 1959 (SCMP 2136, p. 20).

41. "East China Commune Makes Rapid Progress toward Modernization of Agriculture," NCNA Nanking, Dec. 7, 1959 (SCMP 2154, p. 22).

42. "Northeast China Province on the Road to Mechanization," NCNA Harbin Nov. 8, 1959 (SCMP 2136, p. 20).

43. "Agro-technicians and Tractor Drivers Trained in Northwest China Province," NCNA Sian, April 13, 1962 (SCMP 2722, p. 21).

44. Kang Chao estimates that the MTS's owned 10,995 standard tractors at the end of 1958 (*Agricultural Production in Communist China, 1949–1965*, [Madison: University of Wisconsin Press, 1970], p. 109).

tractors were used to open up new stations; from 1957 to 1959, the MTS system expanded from 383 to 553 stations. Perhaps the plan was to sell the equipment to communes after personnel were trained and operations stabilized.

The system of commune management had many problems, especially with maintenance of machinery, and state-operated tractor stations were soon re-established, as will be described in Chapter 7. While the vision of a collective rural group owning and managing its own machinery was not a viable policy in 1958, the idea was implanted for future use.

7. The Ten-Year Plan for Agricultural Mechanization

From 1959 to 1962, China embarked on a major program of farm-tool reform. It was not the frenzied campaign of the Great Leap Forward but an ambitious program for the complete reform of farm tools in one decade. The first stages of the program were successfully carried out. Irrigation was mechanized in some major suburban regions, and tractorization advanced in a few selected places. In a few years, however, the program bogged down and was temporarily abandoned in 1962. What happened? Why could the program succeed only partially?

Three reasons interact to explain the suspension of the program. First, in the aftermath of the mistakes of the Great Leap Forward and the bad weather of 1959–61, the country entered a deep agricultural and economic crisis, which cut off funds for investment in mechanization. Second, administrative problems resulted in popularization of some tools which were technically inappropriate. Third, from an agricultural-economic point of view, many people in China doubted that there was a labor shortage.

In retrospect, this third factor seems especially important. Not until the middle and late 1960's were modern inputs available (chemical fertilizer, high-yielding varieties of rapidly maturing grains, mechanized irrigation, new ways of pest control) to permit further intensification of agriculture through increased multiple cropping and intercropping. When this happened, the peak labor demands went up and the time constraints became more important. The agrotechnical system then required more mechanization. But a program to mechanize agriculture in 1959–61, before the complementary inputs for intensification were available, had little economic logic and fell flat.

Emergence of the Ten-Year Plan

In a broad sense, the Chinese leadership seemed convinced of the appropriateness of the tool-reform campaign of 1958 and with the way it was actually carried out. The Central Committee, meeting in Wuhan in December 1958, reaffirmed the need for tool reform: "It is true that there is a *labor shortage* at present, but the way out must be found in stressing the successful implementation of the *reform of tools* and improvement of labor organization and not in extending working hours."[1]

Plans for 1959, however, suggested subtle but important differences in approach. Gone was the frenzy of the mass campaign. Instead, the plan for tool reform was careful and organized. A *People's Daily* editorial on January 21, 1959, outlined the new program. It asked for careful experimentation and improvement before adopting new tools. Technically qualified personnel were required in this process. The editorial warned: "Sole reliance on the masses will not work."[2] To follow up the decision to improve the organization of tool reform, a National Conference of Directors of Tool Renovation Offices was held in February 1959. The conference decided to select those tools which could save the most labor and to organize local factories to produce them.[3]

Why did tool-reform policy change? Did it reflect inadequacies in the theory or implementation of the mass campaign for tool reform? Did it reflect success of the campaign, permitting new policies to be adopted? Or did it reflect an overall shift in political mood which recognized that mass mobilization would not be able to solve all problems? Actually, all these factors were involved.

In a broad sense, 1959 began with a sorting through of the dramatic changes in organization which had occurred in the fall of 1958. Those changes considered suitable were consolidated. Some of the excesses were discarded. Where policy was too "advanced" for the popular consciousness (for example, Chinese peasants would not accept free distribution of goods or egalitarian leveling among different villages), policy was changed to better suit popular values. Other policies were reaffirmed, including the general organization of the commune and the merger of government offices with units of economic cooperation. The Wuhan Resolution, cited above, was indicative of this process. Thus China was, in a

1. "Resolution on Some Questions Concerning the People's Communes," adopted by the Eighth Central Committee, Sixth Plenary Session, Dec. 10, 1958; available in Robert Bowie and John Fairbank, *Communist China, 1955–1959* (Cambridge: Harvard University Press, 1965), p. 498. Emphasis added.

2. "Speed up the Improvement of Farm Tools," *Jen-min Jih-pao* editorial, Jan. 21, 1959 (SCMP 1951, p. 20).

3. "National Conference of Directors of Tool Renovation Offices Ends," NCNA Peking, Feb. 19, 1959 (SCMP 1964, p. 15).

general way, making marginal adjustments to new policies, and naturally there was some re-evaluation of agricultural tool-reform policy.

One more specific reason for the policy shift has to do with the backyard-furnace campaign in 1958. Some experiments with small village furnaces to smelt iron were reported first in May 1958, and were fairly widespread by June.[4] One of their purposes was to supply materials for agricultural tool reform.[5] At the same time, however, the backyard-furnace campaign made agricultural tool reform even more important. Perhaps 60 to 100 million workers were diverted from agricultural production to make, supply with raw materials, and operate the furnaces.[6] This represented somewhere between 22 and 45 per cent of the rural labor force.[7] Obviously if such a vast number of workers were taken out of farm work at any but the most slack period (and in fact the campaign was not conducted during a slack period), a serious labor shortage would be induced, requiring rapid tool reform and mechanization. This is, of course, the typical pattern of industrial development and agricultural mechanization, but it is usually spread over decades, not months.

By December 1958 it became clear to the Chinese leadership that the backyard furnaces were not working as planned. They were highly inefficient in their use of resources, especially coal, and produced a grade of pig iron with excessive levels of sulphur and phosphorous. It could not be used as iron; nor could it be smelted into steel without expensive steps to remove the impurities. Even smelting it into refined pig iron was difficult.[8] The backyard-furnace campaign had to be dropped, and with it went the potential source of iron and steel for making farm tools in a frantic campaign. Also, much of the labor shortage diminished overnight.

The responsibility (or opportunity?) for generating a new approach to farm-tool reform which would accept the general principles of labor shortage and mass campaign but would integrate the experiences of 1958 and the need for technical competence fell to Mao Tse-tung. In spring

4. M. Gardner Clark, *The Development of China's Steel Industry and Soviet Technical Aid* (Ithaca: Cornell School of Industrial and Labor Relations, 1973), p. 69.

5. Ministry of Agriculture, Sept. 23, 1958; cited by Anna Louise Strong, *The Rise of the People's Communes in China* (New York: Marzani and Munsell, 1960), p. 29.

6. Clark, *Development of China's Steel Industry,* pp. 68–69.

7. Peter Schran estimates the rural labor force in 1958 at 272 million people (*The Development of Chinese Agriculture, 1950–1959* [Urbana: University of Illinois Press, 1969], p. 64). Chinese economists estimated the rural labor force to be 220 million (Feng Chih-kuo, "How to Solve the Problem of Labor Shortage on the Agricultural Front," *Ching-chi Yen-chiu,* no. 3 (March 17), 1959; ECMM 167 p. 18).

8. Clark, *Development of China's Steel Industry,* p. 70.

1959 he projected a ten-year plan to mechanize China's agriculture. On April 29, 1959, Mao wrote to cadres at all levels (from province to work team):

The fundamental way out for agriculture lies in mechanization. Ten years will be needed to achieve this. There will be minor solutions in four years, intermediate ones in seven, and major solutions in ten. This year, next year, the year after and the year after, we will be relying mainly on improved farm tools and semi-mechanized farming implements. Every province, every district, and every county must establish farm tools research stations and concentrate a group of scientific-technological personnel and experienced carpenters and blacksmiths of the rural areas to gather together all kinds of more advanced farm tools from every province, district, and county. They should compare them, experiment with them, and improve them. New types of farm implements must be trial-produced. When they are successfully trial-produced, test them out in the fields. If they are found to be truly effective, then they can be mass produced and widely used. When we speak of mechanization, we must also include mechanized manufacture of chemical fertilizers. It is a matter of great importance to increase chemical fertilizer production year by year.[9]

According to Red Guards, Mao continued his fight for agricultural mechanization at the Lushan Conference of July 1959.[10] Mao insisted upon—and achieved—the establishment of a new ministry, the Ministry of Agricultural Machinery, to give leadership and organization to the entire effort of tool reform and mechanization. It was established on August 26, 1959, and two days later came the first major public announcement of the Ten-Year Plan for Agricultural Mechanization in *People's Daily*.[11] A major exposition of the plan was offered by Po I-po a few weeks later, when he enthusiastically predicted "that in 1969 [i.e., in ten years] or thereabouts, all the land in our country which can be cultivated by ma-

9. Mao's Intraparty Correspondence, April 29, 1959, *Miscellany,* p. 171. Red Guards referred to this text in "History of Struggle," SCMM 633, p. 12. Some sources give the data of this letter as November 29, 1959. See "A Letter to Production Team Leaders", Nov. 29, 1959, included in *Long Live Mao Tse-tung Thought* (n.p., n.d.), CB 891, p. 33, and Jerome Ch'en, ed., *Mao Papers* (London: Oxford University Press, 1970), p. 7. Many of the polemics on agricultural mechanization quote from it. It is interesting, however, that none of the Red Guard groups in agricultural mechanization departments ever cited the last two sentences, which mention the importance of chemical fertilizer.

10. Peking Agricultural Machinery College, East is Red Commune, Criticism and Repudiation Office, "Completely Settle the Heinous Crimes of China's Khrushchev and Company in Undermining Agricultural Mechanization," *Nung-yeh Chi-hsieh Chi-shu,* no. 5 (Aug. 8), 1967 (SCMM 610, p. 18).

11. Cited in Chin Szu-kai and Choa Wing-fai, "The Mechanization of Agriculture," in *Contemporary China, 1962–1964,* ed. E. S. Kirby [selective entry] (Hong Kong: Hong Kong University Press, 1968), pp. 1–9.

chine will be worked, in the main, with mechanical instead of animal traction; and that where irrigation by machine is needed, it will in the main replace human labor.''[12] Po included mechanization of transportation and food processing, as well as complete supply of chemical fertilizer, in this ten-year plan.

Soon after the establishment of the Ministry of Agricultural Machinery, Mao restated his views about the importance of agricultural mechanization. In October, he wrote: ''The mechanization of agriculture is a decisive condition for the development of the 'three-in-one' combination of agriculture, forestry and animal husbandry on a large scale. The Ministry of Agriculture Machinery has been established this year, and it appears that it will not take long to bring the mechanization of agriculture to fruition.''[13]

Policy statements for 1960 indicated continuation of policies stressing mechanization and tool reform. *People's Daily* editorials on January 13 and February 25, 1960, called for tool reform and semimechanization, in both industry and agriculture.[14] In April in a speech to the Second National People's Congress, T'an Chen-lin again endorsed the ten-year, three-stage agricultural mechanization plan.[15]

The three stages of Mao's plan were specified more concretely in *People's Daily* in August 1960.[16]

1. Small-scale solution in four years (1959–63): agriculture, livestock breeding, irrigation and drainage. During this time, mechanization should be achieved in a preliminary way on the outskirts of big cities, market grain growing centers, the major industrial crop growing centers and the major non-staple food growing centers, while the major part of the rural areas concentrate mainly on popularizing semi-mechanized and improved implements.

2. Medium-scale solution in seven years (1964–66). By the end of seven years, mechanization should have materialized over more than half of the rural

12. Po I-po, ''Strive to Carry Out the Great Task of the Transformation of Agriculture,'' *Hung-ch'i*, no. 20 (Oct. 16), 1959 (ECMM 188, p. 1).

13. ''A Letter Concerning the Development of Pig Breeding,'' Oct. 11, 1959, included in *Long Live Mao Tse-tung Thought*, CB 891, p. 33.

14. ''New Development of Tools Innovation in Rural Areas,'' *Jen-min Jih-pao* editorial, Jan. 13, 1960 (SCMP 2180, p. 6); ''Launch an All-People Campaign for Mechanization and Semi-Mechanization of Manual Labor,'' *ibid.*, editorial, Feb. 25, 1960 (SCMP 2212, p. 8).

15. T'an Chen-lin, ''Strive for the Fulfillment, Ahead of Schedule, of the National Program for Agricultural Development,'' report to the second session of the Second National People's Congress, April 6, 1960; NCNA Peking, April 6, 1960 (CB 616, p. 26).

16. Niu Chung-huang, ''On the Technical Transformation of China's Agriculture,'' *Jen-min Jih-pao*, Aug. 26, 1960 (SCMP 2333, p. 1).

areas as a result of the gradual development of the agricultural machinery industry and increased supply of agricultural machines.

3. Large-scale solution in ten years (1966–69). By the end of the ten years, virtually all the countryside as a whole should have mechanization everywhere and also a considerable degree of rural electrification.

The reasons for Mao's surprisingly strong endorsement of mechanization seems to have been a broad political concern with potential cleavages of urban and rural interests. Mao feared that if the benefits of modernization were retained in the urban-industrial centers, the rural people would resent this and become disaffected politically. Mechanization, Mao thought, offered the way to increasing the income of the farmers and reducing their manual toil. It would strengthen the "worker-peasant alliance." Sometime between 1960 and 1962, Mao put this argument together:

> Now, our worker-peasant alliance will go a step further by establishing itself on the basis of mechanization. If there is only cooperativization and communization, but no mechanization, the worker-peasant alliance also cannot be strengthened. . . . Then, on the basis of integration of nationalization [i.e., the eventual transition to total ownership by the state] and mechanization, we will be able to truly consolidate the worker-peasant alliance. Accordingly, the differences between workers and peasants will gradually disappear.[17]

In these observations, Mao was recalling what he had written almost twenty-five years earlier: "The contradiction between the working class and the peasant class in socialist society is resolved by the method of collectivization and mechanization in agriculture."[18]

Implementation

The implementation of the ten-year plan to mechanize agriculture was only partly successful. Unofficially, the plan was dropped in 1962 at the Tenth Plenum. The first of its three stages (mechanization in suburban and commercial crop areas) was, by and large, implemented according to schedule. By 1961, however, the momentum was lost. The thrust shifted from supplying the entire country with machinery to assuring that the machinery already supplied would be properly utilized. This had important political implications; it meant that only some regions would have farm machinery, and only a few specially trained people would operate and

17. Mao Tse-tung, Reading Notes on the Soviet Union's *Political Economics, Miscellany,* p. 256.

18. Mao Tse-tung, "On Contradictions," Aug. 1937, *Selected Works* I, 322.

manage it. Both of these tendencies resulted in political tensions that were revealed during the Cultural Revolution.

Several factors explain the priority attached to the suburban and commercial regions. First was a desire to guarantee food supplies to urban centers. Second, since the urban centers had industrial facilities, they could supply machines, electricity, and experienced mechanics. Third, urban industrial centers could absorb (and needed) whatever rural labor was displaced by machines.

Mechanical Irrigation

In major suburban areas, emphasis was on irrigation and drainage. By 1961, at least ten major mechanical irrigation projects of over one million mou (67,000 ha.) had been completed.[19] The largest of these projects were outside Shanghai, Peking, and Canton. The total mechanically irrigated area increased from 1.13 million ha. in 1961 to 6.6 million ha. in 1965.[20] By 1965 most provinces had some major mechanized irrigation program, but about 78 per cent of the mechanically irrigated land was primarily near the urban centers of Peking, Shanghai, and Canton. Table 7.1 reports data on the growth of mechanized irrigation in various provinces.

One result of this mechanized irrigation was a guarantee of water at almost all times, so that increased multiple cropping became possible. This in turn created labor shortages during the time when one crop was being harvested and the next planted. To alleviate this problem, food processing was mechanized by 1961 in the Shanghai suburbs.[21]

Tractor Distribution

Another important element of the first part of the ten-year plan was tractorization in some regions. Chao Hsüeh, who had written a review of agricultural mechanization in 1957 (cited in Chapter 5), suggested these considerations to guide the distribution of tractors:

In light of the present conditions in our country, we must neither disperse them for use nor go too far in concentrating them for use. If tractors are dispersed

19. "Under the Illumination of the Three Radiant Banners," *Shih-shih Shou-tse*, Jan. 6, 1962 (SCMM 305, p. 12).

20. *Ibid.*; "How China Achieves Good Harvest in 1965," NCNA Peking, Dec. 28, 1965 (SCMP 3609, p. 24).

21. "Rice Threshing Mechanized in Rural Areas around Shanghai," NCNA Shanghai, Dec. 6, 1961 (SCMP 2637, p. 21).

widely in all areas, it would be difficult to look after them, they would easily be damaged, and would not function efficiently. But excessive concentration will imply overall mechanization of agricultural production in certain particular areas. Overall mechanization is bound to give rise to a series of problems relative to labor organization and land planning. Furthermore, farming systems vary from place to place in our country, and suitable farm tools are still wanting for certain kinds of work.

Therefore, the distribution of tractors must be made with reference to the peculiarities of certain areas, while consideration should be given to areas in general. The emphasis with regard to the employment of tractors should be placed on the *plains where the land is large and population is small*, the shortage of manpower and animal power is in evidence, the potentialities for *increased production* are great, the percentage of *commodity production* is high, and communications conditions are favorable.[22]

As a practical matter, only a few geographic regions fit these specifications. The Chinese press reported: "Most of these tractors are being used in China's important grain, cotton, and industrial crop areas in northern and northeastern China, along the Yellow, Huai, Haiho, and Hunho Rivers. The plains are favorable to tractor plowing."[23] "the state has decided to give priority in mechanized farming to Shensi and the other important grain and cotton producing provinces along the Yellow, Huai, and Hai rivers in North China and the Hun River in the Northeast."[24]

Information on the precise distribution of tractors among provinces is fragmentary. What little information there is (summarized in Table 7.2) seems to correspond to the principles stated by Chao. Tractors seem to be used in three basic areas:

1. In Sinkiang, Kansu, and Ninghsia, tractors were used in reclamation-colonizing schemes to establish Chinese strategic presence in Central Asia. This program used demobilized soldiers; tractor operators were often demobilized tank drivers.

2. The most concentrated tractor use was in Northeast China. Heilungkiang used tractors on 20 per cent of the cultivated land by 1961. About 37 per cent of China's tractors were used in the three provinces of Heilungkiang, Liaoning, and Kirin. By 1965, one-third of the land in the

22. Chao Hsüeh, "Several Current Problems of Agricultural Mechanization," *Ta-kung Pao*, Peking, May 15, 1961 (SCMP 2515, p. 23). Emphasis added.

23. "China's Farms Use More Tractors for Spring Sowing," NCNA Peking, April 18, 1961 (SCMP 2482, p. 4).

24. "North China Province to Step up Agricultural Mechanization," NCNA Taiyuan, Feb. 16, 1961 (SCMP 2442, p. 12).

Table 7.1. Mechanical irrigation, 1961–66 (in hectares)

Province	1961	1962	1963	1964	1965	1966
Liaoning	*				*	
Hopei	(100,000)		800,000 [a]	1,300,000 [b]	(1,300,000)	
Shansi	50,000 [c]				(50,000)	
Inner Mongolia					(20,000)	36,000 [d]
Shensi					*	
Ninghsia					11,000 [e]	
Shantung					*	
Kiangsu	(470,000)	1,540,000 [f]	1,600,000 [g]	2,400,000 [h]	(2,400,000)	
Anhwei	266,000 [i]				(280,000)	297,000 [j]
Chekiang	(100,000)	660,000 [k]			(660,000)	
Fukien	(6,000+) [l]				(6,000)	
Hupeh	*	140,000 [m]			(140,000)	
Hunan	*				200,000 [n]	400,000 [o]
Honan					*	
Kwangsi				140,000 [p]	(140,000)	
Kwangtung	*		330,000 [q]	460,000 [r]	(430,000)	388,000 [s]
Kiangsi					(170,000)	173,000 [t]
Szechwan	*				(500,000)	600,000 [u]
Kweichow			10,000 [v]		(10,000)	
Yunnan					13,000 [w]	
Totals accounted for: figures for other provinces unavailable	992,000				6,330,000	
Totals available for 1961 & 1965 only	1,130,000 [x]				6,600,000 [y]	

Note: Figures in parentheses are estimates; asterisks indicate that there is probably substantial mechanical irrigation in the province but no basis for a quantitative estimate.

Sources:

a. "More Electricity for Rural Areas in China's Major Cotton and Wheat Province," NCNA Tientsin, March 8, 1963 (SCMP 2937, p. 19).

b. "Medium and Small Factories Serve Agriculture in North China," NCNA Tientsin, Sept. 7, 1964 (SCMP 3316, p. 12).

c. "China Develops Mechanized Irrigation," NCNA Peking, Dec. 21, 1961 (SCMP 2648, p. 19).

d. "New Pumping Irrigation Project Completed in Inner Mongolia," NCNA Huhehot, Nov. 29, 1966 (SCMP 3833, p. 16).

e. "Pumping Stations Help Transform and Expand Farmland in Moslem Region of China," NCNA Yinchuan, Jan. 2, 1965 (SCMP 3372, p. 19).

f. "China Uses More Pumps for Irrigation and Drainage," NCNA Peking, April 3, 1963 (SCMP 2955, p. 15).

g. "China Uses More Pumps for Irrigation and Drainage," NCNA Peking, April 3, 1963 (SCMP 2955, p. 15).

h. "More Electricity for People's Communes in China," NCNA Peking, April 24, 1964 (SCMP 3208, p. 16). "Power Irrigation Prospers Yangtze Delta," *Chung-kuo hsin-wen,* April 2, 1965, p. 7, reported that over 2,140,000 ha. in the Yangtze Delta were irrigated by electric power (JPRS 30, 311, p. 1).

i. "Farmland Irrigated by Machinery Grows in China," NCNA Peking, Aug. 21, 1961 (SCMP 2567, p. 19).

j. "Construction of Giant Irrigation System in East China Enters Ninth Year," NCNA Hofei, Nov. 13, 1966 (SCMP 3823, p. 28).

k. "Electric Pump Operators and Tractor Drivers Trained for China's Rural Areas," NCNA Peking, Jan. 10, 1963 (SCMP 2898, p. 13). The following year, it was reported that Northern Chekiang, Kashing Administrative District (the region near Shanghai) had 250,000 ha. electrically irrigated. "Mechanized Irrigation and Drainage for Rice Fields in East China," NCNA Peking, Oct. 10, 1963 (SCMP 3080, p. 16).

l. "Water Conservancy Workers in Fukien Prove Fully Effective in Fighting Drought," NCNA Foochow, July 28, 1956 (SCMP 1354, p. 18).

m. "New Electric Pumping Projects in Central China, Yangtze Province, NCNA Wuhan, Jan. 9, 1963 (SCMP 2897, p. 13).

n. "Central China Builds Network of Electric Pumping Stations, NCNA Changsha, Feb. 27, 1965 (SCMP 3408, p. 22).

o. "Giant Pumping Station Network Finished around Tungting Lake by the Yangtze River," NCNA Changsha, July 16, 1966 (SCMP 3742, p. 27).

p. "Big Increase of Water Pumps in South China Region," NCNA Nanning, Dec. 3, 1964 (SCMP 3352, p. 17).

q. "Pearl River Delta Harvests More Rice, Despite Drought," NCNA Canton, Dec. 30, 1963 (SCMP 3133, p. 20).

r. "Fifth Year of Construction on Pearl River Delta Pumping Stations," NCNA Canton, Oct. 23, 1964 (SCMP 3326, p. 19).

s. "Self-Reliance Transforms South China's Pearl River Delta," NCNA Canton, Jan. 10, 1966 (SCMP 3617, p. 24).

t. "Gigantic Rural Electric Power Grid Constructed in East China," NCNA Nanchang, June 4, 1966 (SCMP 3715, p. 20).

u. "China's Leading Rice Producing Province Makes Headway in Technical Transformation of Agriculture," NCNA Chengtu, Oct. 24, 1966 (SCMP 3809, p. 27).

v. "Biggest Pumping Station in SW China," NCNA Kweiyang, Jan. 31, 1963 (SCMP 2912, p. 16).

w. "Electric Pumping Stations Bring Good Harvests to Lake Area in Southwest China," NCNA Kunming, Feb. 5, 1965 (SCMP 3394, p. 24).

x. "Under the Illumination of the Three Radiant Banners," *Shih-shih Shou-tse,* Jan. 6, 1962 (SCMM 305, p. 12).

y. "How China Achieves Good Harvest in 1965," NCNA Peking, Dec. 28, 1965 (SCMP 3609, p. 24).

Table 7.2. Distribution of tractors, 1961–65 (in "standard tractors" units)

Province	1961	1962	1963	1964	1965
Heilungkiang	16,000[a]	20,000[b]	18,000[c]		
Kirin	(3,000)				4,400[d]
Liaoning	(14,000)	16,000[e]			
Hopei	10,300[f]				
Shansi	3,800[g]				
Inner Mongolia	530[h]				
Shensi	(1,000)	1,200[i]			
Kansu					2,300[j]
Ninghsia				1,000[k]	
Sinkiang	3,000[l]				10,000[m]
Shantung	6,900[n]		8,000[o]		
Kiangsu	3,000[p]				
Anhwei	2,800[q]				
Chekiang					
Honan					
Szechwan					
Kwangtung		1,100[r]			
Number located	64,330				
Total[s]	90,000	103,000	115,000	123,000	130,500

Note: Figures in parentheses are estimates.

Sources:

a. *Economic Background* (Hong Kong), no. 751 (1962), p. 51; cited by Kang Chao, *Agricultural Production in Communist China, 1949–1965* (Madison: University of Wisconsin Press, 1970), p. 116.

b. NCNA, Jan. 2, 1963; cited in *ibid.*

c. "Communes in Northeast Begin Mechanizing Farming," NCNA Harbin, Sept. 20, 1963 (SCMP 3067, p. 20).

d. "Success of Mechanized Demonstration Farms in Northeast China," NCNA Changchun, Feb. 18, 1965 (SCMP 3403, p. 22).

e. Estimate by Chao, *Agricultural Production,* p. 116.

f. "More Tractors Used in Northern Part of China for Autumn Plowing," NCNA Peking, Oct. 12, 1961 (SCMP 2601, p. 6).

g. "North China Province to Step up Agricultural Mechanization," NCNA Taiyuan, Feb. 16, 1961 (SCMP 2442, p. 12).

h. "More Tractors for Spring Sowing in China," NCNA Peking, March 19, 1961 (SCMP 2463, p. 16).

i. "Farm Machinery and Equipment for NW China Province," NCNA Sian, Aug. 4, 1962 (SCMP 2796, p. 18).

j. "More Tractors in Northwest China Province," NCNA Lanchow, Sept. 7, 1965 (SCMP 3536, p. 19).

k. "More Farm Machinery in Central China Province," NCNA Wuhen, Aug. 8, 1964 (SCMP 3277, p. 15).

l. "Tractors Are Being Overhauled in Sinkiang," *Jen-min Jih-pao,* Nov. 23, 1961 (SCMP 2635, p. 10).

m. "China on Way to Agricultural Mechanization," NCNA Peking, Sept. 29, 1965 (SCMP 3550, p. 22).

Northeast was machine-cultivated;[25] this extensive mechanization provided commodity grain for urban areas.

3. By 1961, tractors were beginning to be used in the North China Plain, primarily in the plains of the Yellow, Huai, Hai, and Hun rivers. Three provinces in particular received substantial numbers of tractors: Hopei, Shantung, and Shansi. If we assume (as Chinese economists do) that each standard tractor has the potential for servicing 100 hectares of land, then tractors could have cultivated about 10 per cent of these provinces. Suburban Peking got special treatment; by 1963, 50 per cent of the land in the Peking municipality was machine-plowed, and by 1966, 60 percent of the suitable land was machine-plowed.[26]

No provincial data are available for tractors in South and West China. In general, the large tractors first made in China are unsuitable for South China, except in the flat reclamation areas. One thousand tractors were reported in use by tractor stations in Kwangtung's Pearl River Delta.[27] In addition, some tractors were used for experimental and training purposes. Not until 1964 did China begin mass production of small garden tractors suitable for southern paddy cultivation.

The decision in 1961 to concentrate tractors in the North China Plain was so thoroughly carried out that farm implements were reportedly shipped from South China to North China. Altogether, 320,000 items of farm equipment were planned to be exchanged.[28]

While topography and man-land ratios were the most important reasons

25. "China Extends Range of Tractors," NCNA Peking, Jan. 31, 1966 (SCMP 3631, p. 10).

26. "Electric Pump Operators and Tractor Drivers Trained for China's Rural Areas," NCNA Peking, Jan. 10, 1963 (SCMP 2898, p. 13); "Peking and Surrounding Country Districts Develop New Relationship," NCNA Peking, Feb. 7, 1966 (SCMP 3635, p. 22).

27. P. H. M. Jones, "Machines on the Farm," *Far Eastern Economic Review* 45 (1964), 480. Mr. Jones's source is not given.

28. "Farm Machinery Exchange Held in Peking," NCNA Peking, May 16, 1961 (SCMP 2507, p. 23).

---

n. "More Tractors Put to Use in Shantung," *Jen-min Jih-pao,* Nov. 23, 1961 (SCMP 2635, p. 8).

o. "East China Province Mechanizing Farming," NCNA Tsinan, Oct. 22, 1963 (SCMP 3088, p. 10).

p. "Big Farm Machine Industry in East China Province," NCNA Nanking, Nov. 22, 1961 (SCMP 2627, p. 20).

q. "Central China Province Pushes up Mechanized Farming," NCNA Hofei, Nov. 21, 1961 (SCMP 2627, p. 19).

r. Reported by Jones, "Machines on the Farm," p. 480.

s. Chao, *Agricultural Production,* p. 107.

to emphasize mechanization of the North China Plain, it should be noted that the three leading provinces—Hopei, Shantung, and Shansi—were major cotton producers, supplying about 40 per cent of China's cotton.[29] The first section of the ten-year plan gave priority to the "mechanization of commercial crops." Cotton-producing regions probably received tractors in an effort to both raise and stabilize production to assure high efficiency and better planning in the textile industry. It will be recalled that decline in cotton production in 1953–54 was considered very serious by Chinese economists.

Another reason to mechanize was to assist North China in developing agricultural resources to become self-sufficient in grain, and thus to end the need to ship grain north.

In tractorizing certain provinces and suburban regions in North China, the Chinese recognized that it was essential to train managers, operators, and mechanics to use the machinery efficiently. If anything, however, they underestimated the magnitude of this task. By 1961, it was clear that there was little point expanding the role of machinery until the existing system was adequately consolidated with regard to personnel, supply of spare parts, and administration.

To provide personnel for better design, construction, and maintenance of agricultural machinery, specialized education was improved after 1959. Within the new Ministry of Agricultural Machinery a Department of Education, with an "Institutional Education Division," was established to coordinate formal education in agricultural machinery in schools.[30] Training facilities for top-level machinery experts were expanded. Before 1958, thirty-four technical institutes relating to farm machinery had been established. From 1958 to 1960, seventeen new insti-

29. Cotton production in three provinces:

	Cotton production (million tons)	National rank	% of national total
Hopei	.314	1	18
Shantung	.275	2	16
Shansi	.098	7	6
Total	1.687		40

From *Provincial Agricultural Statistics for Communist China* (Ithaca: School of Industrial and Labor Relations, 1969).

30. Institutional Education Division, Department of Education, Ministry of Agricultural Machinery, "Introducing a Few Schools of Agricultural Machinery," *Chung-kuo Nung-yeh Chi-hsieh,* no. 6 (June) 1962, p. 32 (JPRS 43,912, p. 58).

tutes were opened. In 1957 student enrollment was about 5,000; it doubled during the next two years. In 1960 there were 7,600 more students in agricultural mechanization than in 1957.[31]

Little is known about these facilities; two major ones appear to have been in Nanking (formed from farm machine departments in Central University of Nanking and Nanking University) and Peking (the Peking Agricultural Machinery College, which graduated 1,300 technical workers before 1961).[32] Other schools specializing in agricultural machinery were Kirin University of Industry, Chengchiang College of Agricultural Machinery, Loyang College of Agricultural Machinery, Wuhsi School of Agricultural Machinery Manufacturing, Tientsin School of Agricultural Machinery Manufacturing, and Chiamuszu School of Agricultural Machinery Manufacturing.[33] Most of these were attached to agricultural machinery factories.

Substantial additional training of operators, repairmen, and managers was done at the province and local levels. Extensive training programs—involving more than 10,000 people—were reported in Hopei in February 1961.[34] Other programs were reported in early 1961 in Shantung, Shansi, Peking, Honan, and Liaoning—precisely the areas slated for mechanization.[35] Shantung established an agrotechnical institute, an institute for farming mechanization, four vocational secondary schools, and county-level tractor training classes. Shensi set up twenty-three three-year agricultural technical schools with 10,000 students, and conducted short-term training classes in addition. In Liaoning, the Shenyang Agricultural College ran three-month training courses for tractor drivers. Each course enrolled 300 people.[36]

31. "Big Drive to Train More Experts in Mechanization for China's Agriculture," NCNA Peking, Dec. 27, 1960 (SCMP 2410, p. 13).

32. *Ibid.*; "Graduates from Specialty of Agricultural Mechanization of the Institute of Agricultural Mechanization of Peking Rush to Basic-level Units," *Kuang-ming Jih-pao*, Oct. 5, 1961 (SCMP 2603, p. 15).

33. "Introducing a Few Schools of Agricultural Machinery," JPRS 43,912, p. 58.

34. "10,000 More Tractor Drivers and Mechanics in North China," NCNA Tientsin, Feb. 4, 1961 (SCMP 2435, p. 9); "10,000 New Farm Machinery Operators in North China," NCNA Peking, March 1, 1961 (SCMP 2450, p. 8).

35. "Big Program in China for Training Farm Machine Operators," NCNA Peking, Jan. 27, 1961 (SCMP 2430, p. 3); "North China Province to Step up Agricultural Mechanization," NCNA Taiyuan, Feb. 16, 1961 (SCMP 2442, p. 12).

36. "Tractors Increase in Coastal China Province," NCNA Tsinan, May 30, 1961 (SCMP 2510, p. 24); "NW China Agricultural Technical School Graduates State Work," NCNA Sian, Sept. 2, 1961 (SCMP 2574, p. 23); "College in NE China Opens Training Course for Tractor Drivers," NCNA Shenyang, Aug. 7, 1961 (SCMP 2557, p. 17).

During this time more maintenance and repair shops were established. In Shansi, plans were made for 900 such depots. In Shantung, from 1957 to 1961, thirty-four repair and assembly shops, capable of doing complete overhauls on tractors, were established to supplement the original two. Hopei likewise increased the number of repair shops from two in 1957 to fifty in 1961. Every county and district had at least one. Kiangsu established fourteen major repair plants to service agricultural machinery. Honan had seventy factories and workshops and 1,100 local repair centers. Altogether, 330 farm machine repair factories and workshops were reported in Hopei, Peking, Shansi, Shantung, Honan, and Liaoning.[37]

To assist those areas without adequate repair facilities, the Loyang tractor factory sent out 1,000 skilled workers with tools and spare parts in early 1961 to repair tractors and teach drivers and mechanics.[38] To better supply these repair shops and tractors, production of spare parts and implements was stepped up. Special committees were established in some provinces to coordinate industry, trade, and other departments at province, region, and county levels.[39]

Other steps were taken to improve the efficiency of agricultural machinery. In Shansi there were plans to level land, fill gaps, and build adequate roads and bridges for tractors.[40] In Anhwei the tremendous power of defining plots of land was delegated to (or usurped by?) tractor stations:

> The tractor stations are *responsible for the cultivation of land on behalf of the commune,* guaranteeing key-point management, *fixing areas,* and *joining areas*

37. "North China Province to Step up Agricultural Mechanization," NCNA Taiyuan, Feb. 16, 1961 (SCMP 2442, p. 12); More Tractors Put to Use in Shantung," *Jen-min Jih-pao,* Nov. 23, 1961 (SCMP 2635, p. 8); "More Tractors for North China Province," NCNA Tientsin, Dec. 12, 1961 (SCMP 2641, p. 20); "Increased Use of Tractors in Autumn Plowing near Peking," NCNA Peking, Oct. 20, 1961 (SCMP 2606, p. 23); "Big Farm Machine Industry in East China Province," NCNA Nanking, Nov. 22, 1961 (SCMP 2627, p. 20); "Farm Machinery Repair Network Established in Northern China," NCNA Peking, May 23, 1961 (SCMP 2507, p. 23).

38. "China's Biggest Tractor Plant Sends Skilled Workers to Rural People's Communes," NCNA Peking, March 17, 1961 (SCMP 2462, p. 18).

39. "North China Province to Step up Agricultural Mechanization," NCNA Taiyuan, Feb. 16, 1961 (SCMP 2442, p. 12); "Tractors Are Being Overhauled in Sinkiang," *Jen-min Jih-pao,* Nov. 23, 1961 (SCMP 2635, p. 10); "Chinese Machine Building Workers Step up Output for Rural Areas," NCNA Peking, Dec. 8, 1961 (SCMP 2639, p. 12); "China's Machine-Building Industry Helps in Spring Farming Preparation," NCNA Peking, Feb. 6, 1961 (SCMP 2436, p. 10).

40. "North China Province to Step Up Agricultural Mechanization," NCNA Taiyuan, Feb. 16, 1961 (SCMP 2442, p. 12).

together for use of machines. This has further raised the efficiency of the machines. Tractor stations in Fengt'ai, Kuoyang and Hao counties of Fuyang administrative district, after making concentrated use of tractors, have increased the rate of operation of the machine by 37 percent.[41]

There is no way of knowing whether this happened frequently, but the idea received publicity and encouragement.

Another method of strengthening tractor management was to utilize the Young Communist League. In Shantung in 1961, the 14,000 tractor drivers and the 16,000 other workers in tractor stations were mostly young people. Thus it was decided to set up YCL groups at all tractor stations and in all tractor teams to sensitize the young operators and repairmen to a wide variety of issues: servicing the machines, economy of fuel, the importance of enforcing contracts and establishing proper relations with the surrounding farmers, and fulfillment of production tasks.[42]

Managerial Problems

The expansion of tractorization took place in the context of commune ownership and management of tractors. This managerial system, while perhaps sensible in the long run, was very difficult to implement in 1958–60 for a variety of reasons. The result was that China's tractor fleet was not used efficiently, thus undermining the ten-year plan for agricultural mechanization.

Some problems were to be expected; the commune ownership system could not fully resolve contradictions between machine management and agricultural production which had been observed earlier with the machine tractor stations. At the end of 1959, three different systems of ownership were in use in Heilungkiang, each representing a different attempt to resolve the contradictions:

1. Ownership and management by commune. The problem with this system was that "since the crop management brigades were not responsible for the management of the machines, they felt no necessity of setting up any production responsibility systems, and the absence of such a system prevented them from ef-

41. "Tractor Stations Set up in Anhwei to Plow Land for Communes," *Jen-min Jih-pao,* Nov. 23, 1961 (SCMP 2635, p. 9). Emphasis added.

42. "Shantung YCL Provincial Committee Issues Notice on Serious Strengthening of YCL Work on Tractor Stations," *Chung-kuo Ch'ing-nien Pao,* Sept. 24, 1961 (SCMP 2603, p. 10); "Shantung YCL Provincial Committee Calls upon Young Tractor Drivers in the Whole Province to Plow More Land, Carry Out Plowing Properly, Take Good Care of the Tractor, and Save Fuel," *ibid.* (SCMP 2603, p. 12).

fectively raising their efficiency, improving quality of their operations, and reduced their crops."

2. Ownership and management by the brigade. This system facilitated responsibility and efficiency, but made it difficult for the commune to lead in the utilization and planning of production.

3. Ownership by commune, management by brigade. This system represented an attempt to solve the contradictions between keeping the machines in repair and the needs of agricultural production.[43]

To solve analogous problems the management system within the commune began to resemble that of the MTS system. One commune in Heilungkiang applied a "three quota, three fixed, three guarantee" system,[44] meaning;

Three quotas (to be achieved by the commune tractor station)

1. Quota for the task to be accomplished
2. Quota for the fuel oil
3. Quota for the repairs to be performed by the station

Three fixed resources (available to the station)

1. Fixed plots of land
2. Fixed number of machines
3. Fixed number of personnel

Three guarantees (to be given by the station)

1. Guarantee of the time of completion of plowing
2. Guarantee of quality of plowing
3. Guarantee of safety

More serious problems had to do with machine maintenance. At the end of 1958, for example, 20 per cent of the tractors were out of commission and in need of repair, while 40–50 per cent of the irrigation/drainage machines needed repair. At the end of 1960, the figures were 20 per cent for tractors and 20–30 per cent for irrigation/drainage machines. By the end of 1961, 65 per cent of the tractors and 48 per cent of the irrigation/drainage machinery needed repair.[45] These year-end statistics probably make the situation appear worse than it was because they reflect the

43. Ou-yang Ch'in, "Questions Concerning Agricultural Mechanization," *Jenmin Jih-pao,* Dec. 21, 1959 (SCMP 2174, p. 23).

44. Ou-yang Ch'in, "Overall Planning and 'Walking on Two Legs' Essential to Agricultural Mechanization," *Hung-ch'i,* no. 7 (April 1), 1960 (ECMM 209, p. 12).

45. Li Ch'ing-yü, "A Summary of the Conference on Agricultural Mechanization and Electrification," *Nung-yeh Chi-hsieh,* no. 1 (Jan. 15), 1959 (ECMM 161, p. 18); Chao Hsüeh, "Several Current Problems of Agricultural Mechanization," *Ta-kung Pao,* Peking, May 15, 1961 (SCMP 2515, p. 23); Chang Feng-shih, "Produce Suitable Agricultural Machinery to Support Agriculture," *Kung-jen Jih-pao,* Jan. 18, 1962 (JPRS 12,090, p. 11).

accumulated wear on machinery during the entire agricultural season. Statistics are not available on the number of tractors out of service in early spring, when plowing begins. (For the sake of comparison, in 1970 roughly one-fourth of New York City's garbage trucks were out of service at any one time. In 1975 about one-third were out of service; in early 1976 40 per cent were out of service.)[46]

Another problem was the rather high accident rate for personnel involved in tractor operations. Reports from 15 provinces, municipalities, and autonomous regions revealed that 55 operators of commune-managed machines were killed and 100 were injured; 31 operators of other machines were killed and 61 injured.[47] The broad summary of the situation was rather pessimistic. Chao Hsüeh emphasized the maintenance problem in an article published in May 1961: "Owing to the fact that the leading organs slackened management after transfer of tractors to the lower level, a number of grave problems occurred: in particular, damage to engines and injury and loss of lives assumed serious proportions."[48]

There were many reasons underlying the problem of maintenance. The first was, of course, the severe agricultural crisis of 1960, which eliminated agricultural surpluses that could have provided investments for machinery and support services.

The second problem was the very success of the mechanization program. From 1957 to 1961 the total number of tractors in China almost quadrupled (see Table 7.3). Such a rapid expansion naturally presented problems of management and maintenance. Personnel—including operators, maintenance men, and managers—had to be trained very rapidly. Production and distribution of spare parts had to be stepped up enormously. Repair shops had to be set up quickly.

Because over 100 types of tractors and internal combustion engines were in use in China, some domestic, others imported from over a dozen countries, supply of spare parts was difficult at best. The shortage of spare parts and repair facilities was so great that one agricultural mecha-

46. Richard Phalon, "Private Garbage Hauling Is Being Studied by City," *New York Times*, June 6, 1971; "City Trash Pickups at a Critical Stage, Officials Assert," *ibid.*, March 7, 1976.

47. Li, "Summary of the Conference," ECMM 161, p. 18. The period covered by these statistics is not specified.

48. Chao, "Several Current Problems of Agricultural Mechanization," SCMP 2515, p. 23. Similar problems were reported in the Soviet Union when the MTS system sold tractors to the kolkhozy (Lazar Volin, *A Century of Russian Agriculture* [Cambridge: Harvard University Press, 1970], p. 468).

nization expert suggested that for every yuan invested in tractor factories, another yuan be invested in factories to manufacture implements, and two and one-half yuan be invested in spare-parts factories and repair shops.[49]

Complicating matters, the supply system for spare parts was changed in 1959. Before 1959, spare parts and repair services were coordinated by the agricultural departments (local representatives of the Ministry of Agriculture). In 1959, it was decided—not irrationally—that repair services should be organized by industrial departments, and that spare parts should be distributed through normal commercial channels.[50] Red Guards

Table 7.3. Total standard tractors in China, 1957–61

Year	Number of tractors	Index
1957	24,629	100
1958	45,330	184
1959	59,000	240
1960	79,000	320
1961	90,000	366

Source: Chao, *Agricultural Production,* p. 109.

charged that this new repair and supply system was not efficient. Although there is no information to confirm the Red Guard charges, it would seem that even if the new system were better in theory than the pre-1959 method, considerable time might have been required to establish it. Thus it is possible that this administrative change in 1959, although sensible in the long run, magnified the problems of 1959–61.

All these problems were immensely aggravated in the fall of 1960, when the Soviet Union withdrew technical personnel and trade fell off sharply. Before 1960, almost all of China's tractors had been imported from the Soviet Union and Eastern Europe, and many spare parts were imported. One article written later said that problems had been caused by the termination of Soviet aid and trade: "The imperialist blockade and the perfidious action of the modern revisionists in tearing up agreements had

49. Hsiang Nan, "Certain Problems of Agricultural Mechanization," *Jen-min Jih-pao,* Dec. 22, 1962 (SCMP 2900, p. 9).

50. "History of Struggle," SCMM 633, p. 22; Eighth Ministry of Machine Building, United Committee of the Revolutionary Rebels, "57" United Detachment, "Wipe out State Monopoly and Promiote Mechanization on the Basis of Self-Reliance in a Big Way," *Nung-yeh Chi-hsieh Chi-shu,* no. 6 (Sept. 18), 1967 (SCMM 610, p. 12).

caused many difficulties to the building of the farm machinery industry.''[51]

Statistics on Sino-Soviet trade (in Table 7.4) reveal that in 1961 Chinese imports of tractors fell off drastically, to merely 2 per cent of the preceding year. Not until 1963 were substantial imports resumed. The decline in spare parts, however, was not nearly so drastic. Imports in 1961 were 55 percent of those in 1960. This decrease may have caused serious shortages in spare parts, but it did not completely cripple farm machinery. It is not known whether the removal of Soviet technical personnel had a serious impact in this field. The drop in Sino-Soviet trade made it even clearer to the Chinese leaders that China could not mechanize its agriculture by relying on imported equipment and spare parts.

Yet another problem appears to have been the details of implementing decentralization in 1958. Many tractor stations did not adequately prepare communes to receive the tractors. Red Guards charged that machine tractor stations ''very quickly handed over large numbers of tractors to the communes with no regard for the conditions of the latter. In some places, after a meeting was held in the morning, the tractors were driven away in the afternoon.''[52] Moreover, after tractors were sent down, the Agricultural Machinery Bureau of the Ministry of Agriculture abdicated responsibility, held no conferences, and made no suggestions, according to Red Guard groups.[53] How widespread such failures were is not known. Some Red Guard groups charged that this failure in instruction and leadership was conscious sabotage by the ''capitalist roaders'' to discredit the idea of commune ownership and management. Other Red Guards preferred to say that the MTS failures to educate the communes were a result of ''indifference.''[54]

It was also claimed that sending down tractors was merely a technique to reduce government expenses for agricultural mechanization by transferring to the communes the expenses that had been borne by the state in subsidizing tractor stations. Thus proper preparation was irrelevant. Liu Shao-ch'i reportedly argued: ''The State has spent vast sums of money for production of tractors, but in the end it has had to pay out compensation. It had better lay down this burden.''[55] He is also reported to have

51. Hsu Pin-chou, ''Struggle for the Realization of Agricultural Mechanization,'' *Kung-jen Jih-pao,* Oct. 10, 1964 (SCMP 3329, pp. 11–12).

52. ''Wipe Out State Monopoly,'' SCMM 610, p. 11.

53. *Ibid.*, p. 12.

54. ''History of Struggle,'' SCMM 633, p. 22.

55. *Ibid.*

Table 7.4. Soviet Union exports of tractors and spare parts to China, 1955–64 (values expressed in thousands of new roubles)

Export	1955	1956	1957	1958	1959	1960	1961	1962	1963	1964
Tractors (physical units)	1,191	803	68	2,656	941	1,579	33	62	979	688
Value of tractors	6,480	5,300	410	9,680	2,363	2,790	220	93	3,841	4,855
Value of tractor spare parts	1,370	900	1,190	1,540	1,556	1,324	735	855	2,265	1,333

Source: Foreign Trade of USSR (*Statistical Survey*) (Moscow: Ministry of Foreign Trade). The annual volumes have the following JPRS and monthly catalogue numbers: 1955 and 1956 (341:58/17,047); 1957 (526:59/4,973); 1958 (1,087:60/1,914); 1959 (6,220:61/1,031); 1960 (13,349:62/12,040); 1961 (16,086:63/647); 1962 (21,948:64/1,202); 1963 (29,325:65/7,902); 1964 (34,214:66/6,133).

said: "The State has to spend sums of money to run tractors in this way [i.e., MTS]. It takes a sum of money to build the tractors and another sum of money has to be lost in the end. It is better to send them to the grassroots level."[56]

All of these factors were minor contributors to inefficient management of machinery by the communes. The critical element was the fact that in about 1960, in response to the broad economic crisis, the communes were restructured in a manner giving considerable autonomy to the production brigades and teams and reducing the power of the commune-level administration. The sixty-point "Rules and Regulations for the Operation of the Rural People's Communes (Draft)" issued in March 1961 and the revised sixty-point document issued in September 1962 stated clearly that the production team was the basic accounting unit. While the commune retained some managerial functions, it was anticipated that most investments in agricultural inputs would be made by production teams.

This new structure made commune ownership of large agricultural machines virtually impossible. Communes lacked the political and economic strength to make large purchases. Brigades and teams also lacked the scale and financial resources to purchase large machines. In May 1962, a *People's Daily* editorial pointed out that agricultural mechanization policy would have to be geared to taking the production team as the accounting unit.[57] Some time in 1962 (probably November) Teng Tzuhui explicitly spelled out the implications of commune decentralization for tractor ownership and management:

> Since the changeover to take production teams as basic accounting units, capital funds have generally not been sent to the grassroots level. . . . Therefore, the reliance on production teams to buy tractors is definitely beyond their means, and whether or not they have the means to buy garden tractors is also not sure. Tractors are of course principally bought by state tractor stations.[58]

In response to these managerial problems, a state-run tractor station—much like the MTS from the mid-1950's—quietly reappeared. The first indication that policy was being changed was a statement by Liao Lu-yen (released by Red Guards) at a National Conference on Agricultural Work

56. "Wipe Out State Monopoly," SCMM 610, p. 11.
57. "Pay Attention to Maintenance, Repair, and Production of Medium Sized Farm Implements," *Jen-min Jih-pao* editorial, May 20, 1962 (SCMP 2750, p. 11).
58. "Wipe Out State Monopoly," SCMM 610, p. 12; "History of Struggle," SCMM 633, p. 22.

some time in 1961: "Tractors should mainly be operated on the basis of state ownership."[59]

By June 1961, a shift in policy was apparent. A *People's Daily* editorial on June 11, entitled "Properly Manage and Make Better Use of Farm Machinery," implied that agricultural machinery was actually owned by the state, not by the communes: "Steps must be taken to make members of all communes with agricultural machinery understand that agricultural machinery is the precious property of the state and the people."[60]

It is not known whether specific directives were issued calling for the re-establishment of tractor stations at this time, but in December 1961 it was reported that most of the tractors in Hopei were operated by state tractor stations.[61] Tractor stations were re-established in Anhwei at the same time. The Anhwei report received substantial publicity in *People's Daily,* and must be considered an indication of the Central Committee's policy with respect to tractor management at the time.[62]

Other provinces seemed to follow the policy of developing machine tractor stations during 1962. In January, it was reported that Kweichow had 60 stations.[63] In May, reports came from Kiangsi indicating that tractor stations would be developed.[64] Over the summer of 1962, Northeast China switched back to tractor stations; about 400 stations were established in Liaoning, and 484 were organized in Heilungkiang.[65] Stations were established in Shensi and Kwangtung in November, and in Ninghsia in December.[66] By the end of 1962, Red Guards stated, 88 per

59. This sentence is quoted in several articles: Liao Nung-ko, "Knock Down China's Khrushchev and Completely Discredit by Criticism His Line of State Monopoly," *Nung-yeh Chi-hsieh Chi-shu,* no. 2, 1968 (SCMM 620, p. 5); "Wipe Out State Monopoly," SCMM 610, p. 12; "History of Struggle," SCMM 633, p. 22.

60. SCMP 2526, p. 11.

61. "Most Tractors for North China Province," NCNA Tientsin, Dec. 12, 1961 (SCMP 2641, p. 20).

62. "Tractor Stations Set up in Anhwei to Plow Land for Communes," *Jen-min Jih-pao,* Nov. 23, 1961 (SCMP 2635, p. 9).

63. "Tractor Stations for SW China Multi-national Province," NCNA Kweiyang, Jan. 26, 1962 (SCMP 2671, p. 18).

64. Kuo Kuo-sheng, "Development of Potentials of Plowing by Tractors as Seen from the Case of Two Tractor Stations," *Kung-jen Jih-pao,* May 17, 1962 (SCMP 2752, p. 15).

65. "Most Tractors in NE China Province," NCNA Shenyang, Aug. 26, 1962 (SCMP 2811, p. 20); "About 40,000 Agro-mechanical and Technical Personnel in Heilungkiang Province," *Jen-min Jih-pao,* Sept. 26, 1962 (SCMP 2840, p. 10). See also Chang Ch'ing-t'ai, "On the Question of Strengthening the Operation and Management of Tractor Stations," *Chung-kuo Nung-pao,* no. 11 (Nov. 10), 1963 (SCMM 400, pp. 22–33).

66. "Run Tractor Stations Properly," *Shensi Jih-pao,* republished in *Jen-min Jih-pao,* Nov. 2, 1962 (SCMP 2866, p. 13); "Tractor Farming in Pearl River Delta, South China,"

cent of the tractors operated by the communes had been taken away and given to the AMS system.[67]

Red Guard groups gave interesting accounts of how the transfer to state ownership was achieved:

> The capitalist-roaders of the former county committee and the former farming machinery bureau played some tricks. . . .
>
> On the one hand, they appointed the master and accountant of the state tractor station and then went to [the commune] to take over tractors in a bid to achieve a *fait accompli* and force the commune to deliver up their tractors.
>
> On the other hand, they cultivated personal relations, inviting cadres to dinner and "working on their minds," and guided cadres to have a look at the machinery and equipment of the state station.[68]

Another Red Guard group indicated that communes were cut off from essential supplies so that they had no choice but to give up the tractors. "In some localities credits were refused the tractor stations run by communes. They were not supplied with spare parts, and when their machines broke down, they were not repaired. Moreover, these stations were given no concession with respect to the price of fuel."[69] There is no indication in any of these articles that communes were reimbursed for the tractors—a probable source of friction in those cases where the communes had paid for the machines. Red Guards reported some incidents in which communes refused to hand over their tractors.[70]

It seems likely—although there is no documentary evidence—that the centralization of tractors (like the decentralization) was administrative and financial, rather than physical. Commune tractor stations came under state control; few new stations were built in conjunction with the program

---

NCNA Canton, Dec. 17, 1962 (SCMP 2884, p. 18); Wang Ju-ming, "Financial Departments Must Provide Energetic Aid to Agriculture," *Nan-fang Jih-pao,* Nov. 13, 1962 (SCMP 2878, p. 6); "NW China Moslem Autonomous Region Sets up More Tractor Stations," CNA Yinchuan, Dec. 14, 1962 (SCMP 2883, p. 16).

67. Eighth Ministry of Machine Building, Agricultural Machinery Management Bureau, Revolutionary Great Alliance Committee, Mass Criticism Unit, "T'an Chen-lin's Crime of Sabotage against Farm Mechanization Must Be Reckoned With," *Nung-yeh Chi-hsieh Chi-shu,* no. 6 (June 8), 1968 (SCMM 624, p. 5).

68. Revolutionary Leading Team, Hsinchow Farming Machinery Bureau, and Hsinchow New Farming Machinery Revolutionary Rebel Field Corps, "Farm Mechanization Promoted by Communes in Hupeh's Hsinchow Hsien along the Revolutionary Route Charted by Chairman Mao," *Nung-yeh Chi-hsieh Chi-shu,* no. 9 (Sept.) 1968 (SCMM 630, p. 31).

69. "History of Struggle," SCMM 633, p. 23.

70. "Farm Mechanization Promoted by Communes in Hupeh's Hsinchow Hsien," SCMM 630, p. 31.

and machinery was probably no more centralized in physical terms. However, new stations were established as the number of available tractors grew.)

A major document entitled "Suggestions for Readjustment and Improvement of the Work of Tractor Stations," released in November 1962 in conjunction with the Sixth National Agricultural Machinery Conference, summed up the problems with the commune management system:

> [Communes] cannot afford [tractors], nor [can they] operate tractors properly and cannot achieve mechanization independently.
>
> Even when ownership and operation are placed in the hands of the communes, we can't deny the fact that it is still the State which supplies the money, and operating losses are still made up for by State subsidy.
>
> After many tractor stations were turned over to the communes for ownership and operation, the original set of rules and regulations were set aside, resulting in low rate of utilization and serious damage to machines.
>
> When the people's communes take the production team as the accounting unit, operation of tractors by the communes the teams becomes even more difficult.[71]

This document, sent down to local levels, was considered a directive to establish state control over tractor stations, and presumably set up criteria and standards for the new stations. It sealed the fate—at least for a few years—of the experiment in commune ownership.

Thus for a variety of reasons—some related directly to the needs of agricultural mechanization, most arising from major changes in the structure of the rural economy—commune ownership and management of large agricultural machinery was impossible by 1961 and 1962. The actual process of deciding to re-establish the previous form of tractor stations took perhaps a year. Another year was required to collect the tractors and set up a new administrative structure.

While it is clear that the decentralization policy of 1958–61 failed, it is not clear that it ever was given an adequate chance to succeed. It had been carried out quickly, without adequate preparation, at an unfortunate time. Many Chinese leaders refused to believe that the experiences of these years proved that communes could never own and manage agricultural machinery, and advocated this policy again in the late 1960's.

The re-establishment of state-run tractor stations chipped away at the theoretical underpinnings of the ten-year plan for agricultural mechanization. Mechanization was again an element of the state bureaucracy; it was

71. "History of Struggle," SCMM 633, pp. 22–23. Some of the quotations are also available in "T'an Chen-lin's Crime of Sabotage," SCMM 624, p. 6.

no longer an integral aspect of the rural community. Moreover, it was clear that more resources—administration, spare parts, technical personnel—would have to be concentrated in the regions that had the first tractors to assure efficient utilization. Plans to expand the use of tractors beyond the few regions would have to be delayed.

Attempts to Popularize Intermediate Technology

Although the ten-year plan was fairly successfully implemented in major suburban regions, the efforts to popularize improved tools in other regions bogged down. The blame cannot be placed on lack of administrative support. If anything, there was too much encouragement from the central administration. Within the Ministry of Agricultural Machinery a Farm Tool Improvement Bureau was set up that included a Mass Movement Section. The bureau presumably encouraged reform of farm tools around the country. There was, also, a National Office for Farm Tool Reform, which had provincial, district, and county counterparts, probably under the Ministry of Agriculture.[72]

Table 7.5. Farm tool experiment stations, 1959–62

Year	Provincial level	Special district level	County level	Total
1959[a]	26	75	543	644
1960[b]	—110—		n.a.	n.a.
1961[c]	—101—		441	542
1961[d]	25+	75	500+	600+

Sources:
a. "History of Struggle," SCMM 633, p. 15.
b. "Speed Up the Mechanization of Agriculture," *Jen-min Jih-pao* editorial, Sept. 22, 1960 (SCMP 2355, p. 3).
c. "History of Struggle," SCMM 633, p. 20.
d. "New Achievements by China's Agricultural Machine Industry," *Shih-shih Shou-ts'e,* no. 3–4 (Feb. 17), 1962 (SCMM 315, p. 21).

In accordance with the Chengtu decisions and the various editorials, hundreds of agricultural mechanization experimental stations were established at the province, special district, and county levels. Table 7.5 summarizes available information on these stations. Consistent with these plans, investment in local industry (to manufacture improved tools) al-

72. "History of Struggle," SCMM 633, p. 20.

most tripled from 1958 to 1960.[73] This is shown in Table 7.6, which also shows the sharp decline in investment in local factories very soon after 1960.

Indeed, local industry took such a large share of the agricultural machinery budget that central ministry factories received in 1960 only 59 per cent of what they had received in 1958.

As a result of these policies, the production of semimechanized implements showed a sharp increase in 1960 and a high level of production in 1961 (see Table 7.7).

Table 7.6. Investment in agricultural machinery industry, 1958–66

Year	Index of total investment	% of total investment which went to local industry	Value of central investment (as % of 1958)	Value of local investment (as % of 1958)
1958	100			
1959	178			
1960	378	83	59	289
1961	n.a.			
1962	42			
1963	56	23	44	13
1964	77			
1965	n.a.			
1966	n.a.	19		

Source: Computed from "History of Struggle," SCMM 633, pp. 14, 19, 24.

These data on investment and production suggest that the emphasis on farm tool improvement lost steam after 1961. Investments in local industry dropped off sharply, and were only 5 per cent in 1963 of what they had been in 1960. Red Guards charged that half of the farm machinery plants were closed and two-thirds of the workers fired, and that local industries were especially hard hit.[74] Probably hardest hit were those facilities hastily created in 1960. The Hsünte County Agricultural Machinery Plant in Kwangtung, for example, reported its labor force slashed in 1960 from 1,000 to 300.[75] Nationwide, production fell off by about 50 per cent (see Table 7.7).

73. Policy toward local industry before this period and the transition to this new policy are described by Carl Riskin, "Small Industry and the Chinese Model of Development," *China Quarterly*, no. 46 (April–June 1971), 245–273.

74. "History of Struggle," SCMM 633, p. 19.

75. Field Team of the Economic Commission of the Central–South China Bureau of the Chinese Communist Party Central Committee, "Operate Hsien-Level Agricultural Ma-

Table 7.7. Estimated production of semimechanized and improved farm tools, 1953–64

Year	No. of tools (millions)	Year	No. of tools (millions)
1953	.74	1959	n.a.
1954	.91	1960	7.37+
1955	1.62	1961	5.07+
1956	4.89	1962	3.77−
1957	2.94	1963	2.77−
1958	n.a.	1964	3. +

Sources:
1953–57: See Appendix 1. The estimates are crude, but of the correct order of magnitude. The total for this period is known with more accuracy than the individual yearly figures.
1960–63: See Appendix 2.
1964: "Carry out the Agricultural Machinery Work in Revolutionary Spirit in 1965," *Chung-kuo Nung-yeh Chi-hsieh,* no. 1 (Jan.) 1965, pp. 4–6 (JPRS 4,284, p. 1).

Administrative changes were also made to de-emphasize farm-tool reform. The Farm Tool Improvement Bureau, which was originally supposed to encourage the renovation of tools, was converted into the Improved Farm Tool Bureau, to oversee manufacture. Its Mass Movement Section was abolished, and eventually the bureau was also dissolved. In 1961, according to Red Guard charges, the Ministry of Agriculture closed its National Office for Farm Tool Reform and all parallel offices at provincial, district, and county levels.[76] Most of the 500 to 600 tool research institutes were also closed. By 1965, only 92 were left.[77]

Why did this ambitious program to popularize improved farm tools bog down after 1961? The most important reason was the fact that China entered a severe agricultural crisis in 1960, and surpluses were not available to satisfy the food and raw-material needs of industry and its infrastructure. Shortages of raw materials and transportation bottlenecks prevented full utilization of existing tractor factories. Investment funds had to be sharply reduced.

Another reason was administrative. Mechanisms had not yet been found to prevent the Chinese bureaucracy from settling on a "miracle tool" and popularizing it widely before its technical suitability had been

chinery Plants Properly to Support Agricultural Production," *Chung-kuo Nung-yeh Chi-hsieh,* no. 4 (April), 1964, pp. 7–8 (JPRS 42,484, p. 13).

76. "History of Struggle," SCMM 633, p. 20.
77. *Ibid.*, p. 21.

adequately tested. In 1960, the "miracle tool" was the rice transplanter. The first successful transplanter was reported in April 1956 by the Central China Agricultural Research Institute. Additional experimental models were made, and in 1958 a conference selected five for mass production. By spring 1959, 200,000 transplanters were being manufactured in Szechwan, Kwangtung, Hunan, Anhwei, and Kiangsi. A year later, in March 1960, it was claimed that 860,000 had been produced, and by July 1960 the claims rose to 2.3 million transplanters produced in thirteen southern provinces. Another report was that 4.5 million transplanters were being put into use, and that these would be used on 70 per cent of China's rice fields.[78] This probably was the government's goal; it certainly was not achieved. The rapid popularization of the rice transplanter was misguided. Red Guards later criticized T'an Chen-lin, then politbureau specialist in agriculture, for popularizing this transplanter prematurely: "In 1960, he blindly decided to manufacture two million paddy rice transplanters. As no extensive experiments and no detailed designing was done, the machines were useless and a considerable financial loss was sustained."[79]

The extensive production and distribution of rice transplanters from March to July 1960 coincided with a broader political current, which had manifestations in rural, urban, industrial, and foreign policy. In rural policy, the National Program for Agricultural Development, 1956–1967, was approved by the National People's Congress on April 10, 1960.[80] A few days earlier, T'an Chen-lin, speaking to the NPC on the role of communes, suggested activities which had been previously criticized: backyard furnaces and community kitchen facilities. It was in this speech that Mao's personal support for the Ten-Year Agricultural Mechanization Plan was revealed.[81] With regard to urban policy, there was heightened inter-

78. "Rice Transplanting Machine Designed," NCNA Wuhan, April 25, 1956 (SCMP 1278, p. 11); "Conference Selects Five Rice Planting Machines," NCNA Nanking, Sept. 17, 1958 (SCMP 1858, p. 37); "Rice Transplanters Made in China's Major Rice Growing Areas," NCNA Peking, March 30, 1959 (SCMP 1987, p. 20); "Rice Transplanters in Mass Production in China," NCNA Peking, March 9, 1960 (SCMP 2216, p. 11); "Chinese Peasants Make New Tools for Rice Fields," NCNA Peking, July 20, 1960 (SCMP 2304, p. 4); "Rice Transplanters To Be Used on 70 Percent of China's Rice Fields This Year," NCNA Peking, April 14, 1960 (SCMP 2242, p. 17).

79. "History of Struggle," SCMM 633, p. 16.

80. "National Agricultural Development Program, 1956–57," NCNA Peking, April 11, 1960 (CB 616, pp. 1–17).

81. "Full Text of Vice-Premier T'an Chen-lin's Report on Program for Agricultural Development," NCNA Peking, April 6, 1960 (CB 616, p. 22).

est in urban communes during the period of roughly March to July 1960.[82] As for industrial policy, it is known that Mao's vision of industrial management was encapsulated in the Anshan Iron and Steel Company Constitution, released on March 22, 1960.[83] Finally, with regard to foreign policy, in April 1960, in conjunction with the ninetieth anniversary of Lenin's birth, important statements were published, including *Long Live Leninism*. While the broad thrust of these statements was to attack revisionism in the international communist movement, in one of them the Great Leap strategy was reaffirmed: "As Chinese Communists put it, it is possible to advance at leap-forward speed."[84] While little is known about this political current, it would appear that agricultural mechanization policy (this time symbolized by the rice transplanter) was an integral element.

It is appropriate to consider the erroneous and premature popularization of the rice transplanter in a broader context. Throughout the 1950's, a persistent problem in China with regard to diffusion of technology was the frequent popularization of inappropriate items. In agricultural technology, the two-wheel plow, the cable-drawn plow, ball bearings, and the rice transplanter (as well as techniques like close planting and deep plowing) were rapidly popularized without local testing. The higher levels of the political system frantically sought a "miracle tool" which would solve the problems of economic production. In all cases, the tools were basically good, but had not been developed to the point where large-scale diffusion was appropriate. After central authorities selected a particular new tool for mass distribution, lower-level officials knew that concentrating their energies on that tool would be politically safe and would satisfy their superiors. They tended to do so without regard to technical reality until months after the damage was done.

Mao Tse-tung had warned against this tendency among lower-level officials, which he labeled "Left adventurism" in early 1956:

> The targets and plans of all departments must be based on practicability. . . . To carry out something that is impractical and is not fully justified is called blindness

82. Janet Salaff, "The Urban Communes and Anti-City Experiment in Communist China," *China Quarterly*, no. 29 (Jan.–March 1967), pp. 82–10; Ezra Vogel, *Canton under Communism* (Cambridge: Harvard University Press, 1970), p. 267.

83. Editorial Boards of *Wen-hui Pao, Chieh-fang Jih-pao, Party Branch Life*, "Two Diametrically Opposed Lines in Building the Economy," *Jen-min Jih-pao*, Aug. 25, 1967 (SCMP 4012, p. 17).

84. "Forward along the Path of the Great Lenin!" *Jen-min Jih-pao* editorial, April 22, 1960; available in *Long Live Leninism* (Peking: Foreign Languages Press, 1960), p. 61.

or "Left" adventurism. At the present time, this is not the principal tendency in the whole Party, but it is already discernible. Some comrades are not as sober as they used to be and they dare not seek truth from facts. The reason is that the label of Right conservatism or opportunism is not pleasant to hear.[85]

It is particularly important to note that this "Left adventurism" was manifested both in 1956, when China was imitating the Soviet Stalinist pattern of strong, centralized ministries, and during and soon after the Great Leap Forward, when China was experimenting with decentralization and vesting increased authority in the Communist Party. Even the major shake-up of 1958, designed partly to combat the tendencies toward bureaucracy, could not prevent adoption of miracle tools.

While the main reason for the persistent popularization of inappropriate techniques is probably what Mao discerned as fear of being labeled "Right opportünist," it is not inconceivable that some of the problems stem from deliberate sabotage within the bureaucracy. Those opposed to mechanization and semimechanization could have deliberately encouraged irrational policies so that the program could be criticized more thoroughly. Such a tactic is commonly called in China "Waving the red flag to oppose the red flag."

In retrospect, this should not be surprising, considering that China's history and culture revolve around the idea of bureaucracy. For thousands of years, Chinese have been taking examinations to obtain secure, high-status urban jobs. They have been protecting their positions and fighting political battles with bureaucratic guile for generations. It is inconceivable that this dimension of China's political culture would disappear in just a decade. The persistent recurrence of premature popularization of unsound machines serves to highlight the potential conflict between bureaucratic politics and economic growth.

A third reason for halting the farm-tool improvement campaign was that a great many people in China simply did not believe that lack of efficient farm tools constituted a major bottleneck in agricultural production because they felt there was no labor shortage. Red Guards claimed that Li Ching-ch'uan, Party leader of Szechwan, held this view, as did leaders in Chekiang and other areas.[86] T'an Chi-lung, who had been governor of Shantung from 1958 to 1963 and first secretary of the Shantung Provin-

85. Mao Tse-tung, "Talk on the Question of Intellectuals at a Meeting Convened by the Central Committee," Jan. 20, 1956, *Long Live Mao Tse-tung Thought,* CB 891, p. 21.
86. "History of Struggle," SCMM 633, p. 17.

cial CP, was quoted as saying at an unspecified time, but probably in 1961:

> Mechanization, while it can contribute to an increase of overall output, cannot increase yield per unit area. On the other hand, labor would be idle and cannot be used. . . . The cost of irrigation by machines is high and while it can contribute to an increase of overall production, it cannot raise per unit area yield. . . . Water pumps are not so good as water wheels; water wheels are not so good as "Yinyang" cans [for lifting water]; "Yinyang" cans are not so good as Lulotow [another kind of water receptacle].[87]

Bureaucratic rivalries also resulted in some opposition to agricultural mechanization. Some officials in the industrial departments feared that mechanization would take investment funds away from heavy industry. An article in *Red Flag* about agricultural mechanization in Heilungkiang drew attention to this problem:

> All departments, industries, finance and trade organizations and cultural and educational enterprises must actively support agriculture. Industrial departments in particular must treat the development of agricultural mechanization as a glorious task to be accomplished, and take the initiative in strengthening this field of work.
>
> But there are a number of our comrades who lack a clear understanding of this point, believing that agricultural mechanization is something with which they are not immediately concerned. *Some of these comrades even fear that any support given to agricultural mechanization would weaken their own strength, and hamstring the successful operation of the departments to which they are attached.* Swayed by this thinking, they are not enthusiastic as they ought to be with the result that questions which could have been straightened out are not solved.[88]

This sort of bureaucratic rivalry was particularly intense at this time because of the deep financial crisis, which reduced all investment funds.

Some resistance to semimechanization came from rural cadres who argued that full mechanization would come soon, and that therefore semimechanization was at best a waste of time and energy. It might even be harmful, because one criterion for allocating tractors was that the receiving areas not have enough animal power. To develop animal husbandry and semimechanization, then, might mean being bypassed when it came to tractor distribution. This contradiction was noted in Heilungkiang:

87. *Ibid.* These statements are particularly interesting in view of T'an's political career. His rise in Shantung had been accompanied by strong verbal support of the Great Leap Forward. However, he was criticized during the Cultural Revolution and was not listed as a member of the Provisional Revolutionary Committee of Shantung established February 23, 1967. Interestingly enough, he was elected an alternate to the Ninth Central Committee and was present at national day rallies in 1970.

88. Ou-yang, "Overall Planning," ECMM 209, p. 11. Emphasis added.

In the course of promoting mechanization in Heilungkiang Province, cases have been found where some of the areas boycott the use of draft animals and are not enthusiastic in improving the farm tools currently in use. Responsible personnel in those areas feel that inasmuch as mechanization is going to take place, "draft animals will not be of any use, and farm tools, even if they are improved, will not serve any useful purpose." Some of them even went to the length of advocating; "Horses are to be retired from service and farm tools are to be sent to the museum." They suggested the immediate "severance of all relations" with plows and sickles. Some of the people's communes, though having only a very small number of machines, began getting rid of the draft animals and farm tools they originally had after they received tractors.[89]

A similar problem was noted in Chekiang: "There are also some people who want to use modern methods to the exclusion of native methods and want large size to the exclusion of medium size and small size. When they want anything they ask the higher levels for it. They are waiting for mechanization to come of its own accord."[90]

Others opposed mechanization because it could not aid the immediate agricultural crisis. In Chekiang it was charged that "some people erroneously think that the agro-technical reform is 'water in a far away place' which cannot be used to put out a 'nearby fire.' "[91]

Some complained about the "mass movement" aspects of the tool reform campaign. Sung Wei-ching, writing in *Red Flag* on August 1, 1960, denounced this view:

When the [tool innovation] movement first began, it met with ideological hindrance and opposition to the launching of large scale mass movement and the placing of reliance upon the masses for them to effect the reform by their own efforts. The conservative viewpoint held by the rightists is that creation and invention is the business of the technical personnel and is something beyond the workers and peasants; that what has never been tried by our predecessors cannot be accomplished by us; and that mechanization can be achieved just by importing machinery and samples from abroad.[92]

There was also an element of status and pride. Some rural cadres apparently thought it beneath their dignity to show interest in old-fashioned tools: "There are people who look down upon tool reform. They think

89. *Ibid.*, p. 17.
90. Tai Meng, "Speed up the Agrotechnical Reform," *Ch'iu-shih*, no. 3 (March 15), 1960 (SCMM 215, p. 21).
91. *Ibid.*
92. Sung Wei-ching, "Accelerate the Agro-technical Reform, Deepen the Farming Tools Innovation Movement," *Hung-ch'i*, no. 15 (Aug. 1), 1960 (SCMM 228, p. 18).

that Western things are better than native things, and big things are better than small things. They think that tool reform is not worth carrying out, and that what is worth carrying out is agricultural mechanization.''[93]

Finally, there was the factor of finances: ''Some [comrades] feel that agricultural machinery is high in cost and complicated in technique and that the peasants can neither afford it nor operate it. They maintain that since the state is to be relied upon for the realization of agricultural mechanization, it is not necessary to mobilize the masses or launch any mass movement.''[94]

There were, undoubtedly, some who supported mechanization but felt that the ten-year plan was premature. About this time, the Chinese made some estimates about what would be required to mechanize Chinese agriculture. The conclusions of the reports (summarized in Table 7.8), not published until after the Tenth Plenum, certainly dampened any hopes of completing mechanization in ten years. Liu Jih-hsin described the total needs as ''astonishingly large.''[95] Such a phrase was not out of place, especially with regard to tractors. In 1960, China's production was almost 25,000 standard tractors.[96] At that rate, it would require 40 years to produce the required one million. Of course production could rise, but depreciation would have to be added to the needs. These computations must have reminded the Chinese that it was impossible to mechanize agriculture in just ten years.

Machines and implements had still not been designed for various specific needs, and an infrastructure to maintain machines was not yet strong. Thus some advocates of mechanization thought higher priority should be placed on research, development, education, and establishment of spare-parts factories and repair shops. The First National Conference on Agricultural Machinery, meeting in Peking in June and July 1962, emphasized that research into agricultural machinery was lagging behind. Research work was called a ''key factor''; it was, however, ''extremely complicated.'' The conference concluded that research should emphasize (1) tractor-drawn implements, (2) semimechanized tools, (3) standardization of tractors, and (4) improvement of manufacturing and repair of

93. Li Yen-min, ''Tool Reform in Rural Areas,'' *Hung-ch'i,* no. 15 (Aug. 1), 1959 (ECMM 181, p. 13).

94. Ou-yang, ''Overall Planning,'' to ECMM 209, p. 11.

95. Liu Jih-hsin, ''Exploration of a Few Problems Concerning Mechanization of Our Agriculture,'' *Jen-min Jih-pao,* June 20, 1963 (SCMP 3021, p. 3).

96. Kang Chao, *Agricultural Production in Communist China, 1949–1965* (Madison: University of Wisconsin Press, 1970), p. 107.

Table 7.8. Estimates of equipment required to mechanize China's agriculture

Estimator	Machine	Density*	Relevant area (million ha.)	Total need
Hsiang Nan[a]	Tractors	1 st. tractor (15 h.p.) for 100 ha.	100	1.0–1.2 million st. tractors
	Combine-harvesters	1 unit for 300 ha.	100	300,000–350,000 units
	Trucks	1 unit for 300 ha.	100	300,000–350,000 units
	Irrigation/drainage equipment	15 h.p. pump for 13 ha.	53	17.5–20 million h.p.
Liu Jih-hsin[b]	Mechanical equipment	800 h.p. for 10,000 mou	107	130 million h.p.
	Tractors	1 st. tractor (15 h.p.) for 100 ha.	80	800,000 st. tractors
	Trucks	1 unit for 267 ha.	107	400,000 units
	Irrigation/drainage equipment	1 h.p. for 2.67 ha.	53	20 million h.p.
	Chemical fertilizer	225 kg./ha. (gross weight)	107	24 million tons (gross weight)
	Electricity	75 units/ha.	107	8 billion units

Sources:

a. Hsiang Nan, "Certain Problems of Agricultural Mechanization," *Jen-min Jih-pao,* Dec. 22, 1962 (SCMP 2900, p. 5).

b. Liu Jih-hsin, "Exploration of a Few Problems Concerning Mechanization of Our Agriculture," *Jen-min Jih-pao,* June 20, 1963 (SCMP 3021, p. 3).

*Density based on comparisons with USSR, U.S., and Japan. In general, the recommended densities are modest.

machines. The experts seemed to be saying that the rapid expansion of agricultural machinery was premature for technical reasons.[97]

Review of the Ten-Year Plan

While the mechanization program from 1958 to 1962 had some successes, it came nowhere near fulfilling the goals established by the Chengtu Conference in March 1958 or by the ten-year plan of 1959. The plan had envisaged the initial development of mechanization in suburban areas and commercial crop (cotton) areas—a goal that was substantially achieved—and its rapid expansion into the rest of China, which was not. Thus, rather than unifying agriculture in China, the partial achievement of the plan increased (temporarily at least) the differences between different localities.

Likewise, the theoretical and institutional bases of the mechanization program were severely undercut. A basic assumption in 1958 and 1959 was that China was suffering a labor shortage which required mechanization. By 1961 and 1962, it appears that the dominant belief was to the contrary, and that mechanization and tool reform were not considered very important. The Chengtu Conference had recommended financing and management of agricultural machinery by the units of production—cooperatives or communes. This was tried and found unsatisfactory because of problems in maintenance. By 1962, state-operated agricultural machine stations had been established to assure efficient utilization of machinery. The Chengtu Conference had urged emphasis on small-sized farm machines and semimechanization, but the premature and extensive distribution of cable-drawn plows and rice transplanters undercut belief in the utility of semimechanization. Locally based small-farm-tool factories had been suggested by the Chengtu decision. By 1962, many of these had been closed, in the interest of assuring quality and economy.

Does the failure to implement fully the Chengtu decisions and the ten-year plan thus imply that the overall vision of rapid decentralized mechanization is irrational? No, because the failure of the mechanization plan had many causes: the agricultural crisis after 1960; the sharp reduction in trade with the Soviet Union, the low level of production of farm

97. "The Research Work in Agricultural Machinery Should Be Ahead of Agricultural Mechanization," *Chung-kuo Nung-yeh Chi-hsieh,* no. 9 (Sept.), 1962, pp. 2–4 (JPRS 43,912, p. 60); also available as "Scientific Research on Agricultural Machinery Must Precede Farm Mechanization," SCMM 336, p. 14.

machinery, the inappropriate designs of equipment, poor maintenance, and overenthusiastic popularization of inappropriate machines.

Perhaps in an overall sense the program was premature. Many Chinese questioned whether a labor shortage, which would require rapid mechanization, really did exist. The ten-year plan was actually implemented primarily in suburban regions, where the urban economies add to the demand on farm labor. Not until complementary supplies of fertilizer, fast-maturing, fertilizer-responsive varieties of rice, mechanical irrigation, and improved pest-control systems were available in the middle and late 1960's was intensification of agriculture really feasible. Then the increased multiple cropping and intercropping did increase the demand for labor and impose even more serious time constraints on the cropping pattern. Mechanization became sensible.

Important questions concerning the rationality of the political process in China are raised by the fact that such a premature program was attempted. The source of the problem appears to be that the agricultural mechanization program was conceived primarily to encourage broad social transformation. The Chengtu Conference saw mechanization as part of a transformation of rural China, involving the merger of small agricultural producer cooperatives into larger units, the development of rural industry, and the establishment of decentralized management. The precise impact of mechanization on agriculture was of secondary importance. Likewise, the Ten-Year Plan for Agricultural Mechanization, advanced by Mao in April 1959, was proposed at a time when the policies of the Great Leap and of communes were coming under attack. The attack escalated until the Lushan Conference in July 1959, when P'eng Teh-huai presented a broad critique of Mao's policy and leadership. It seems probable that Mao proposed a massive plan for agricultural mechanization in an attempt to sidestep criticism, to strengthen the collective economy, and to improve living standards. Unfortunately, with political effects receiving so much attention, the plan's technical details and problems were given inadequate attention.

In summary, it appears that before 1962 the Chinese were not only aware of the social implications of agricultural mechanization, but tended to be entranced by them, to the extent that the technical problems of mechanization were ignored—until they became unavoidable.

8. The Tenth Plenum and China's Green Revolution

The Tenth Plenum of the Eighth Central Committee of the Communist Party of China, meeting in Peking September 24–27, 1962, marked a major watershed in China's agricultural development strategy. The Tenth Plenum made two crucial decisions concerning agriculture. First, it placed national *priority* on the agricultural sector. No longer would heavy industry be the most important component of economic growth. Second, it decided that agriculture would undergo a *technical transformation*. The socialist transformation of agriculture was considered completed. The communiqué of the Tenth Plenum summarized these decisions:

It is necessary to mobilize and concentrate the strength of the whole Party and the whole nation in an active way to give agriculture and the collective economy of the people's communes every possible material, technical and financial aid as well as aid in the field of leadership and personnel, and to *bring about the technical transformation of agriculture,* stage by stage in a manner suited to local conditions.[1]

The new policy was summed up in the slogan "Take agriculture as the foundation, and industry as the leading factor," which appeared for the first time[2] in the communiqué: "The urgent task facing the people of our country at present is to carry through the general policy of developing the national economy with *agriculture as the foundation and industry as the leading factor*."[3]

These decisions were crucial because the basic thrust of policies in the rural sector had previously been toward changing the pattern of social relations from a private-property system to a collective system, and then

1. "Communiqué of the 10th Plenary Session of the 8th Central Committee of the Chinese Communist Party," NCNA Peking, Sept. 28, 1962 (CB 691, p. 4). Emphasis added.

2. The notion had been around for a few years and had, in fact, been expressed by Stalin in 1926 (*Works* 8:301), but the slogan took its final form in 1962.

3. CB 691, p. 4. Emphasis added.

finding a suitable collective system. Now the emphasis was on technological transformation. A Chinese economist pointed out the shift in priorities: "The Tenth Plenary Session of the Eighth Party Central Committee has decided to place the technical transformation of agriculture on the main agenda of economic construction. . . . The decision of the Tenth Plenum means that in the field of agriculture the main emphasis has been shifted from social transformation to technical transformation."[4]

The emphasis was now on increasing yields and production through intensive utilization of resources, through diversification in cropping systems, and through seed and fertilizer improvements.[5] This had major implications for agricultural mechanization policy. No longer was mechanization seen as the core of agrotechnical transformation and as a spark for rural social transformation. Rather, mechanization was to be simply one of many modern techniques that could help agricultural production, especially if it were used carefully and sparingly. Thus, although mechanization was not expanded much during the early 1960's, important advances were made in research, development, and production of agricultural machinery, and in training of personnel and development of management systems. It was a period of consolidation.

The new emphasis on agriculture, on technical change, and on productivity was not, however, devoid of social content. The Tenth Plenum also admonished, "Never forget class struggle." This slogan stimulated political movements in the following decade and became a major watchword in 1976 in the criticism of Teng Hsiao-p'ing, who was charged with favoring economic development without concern for revolutionary values.

Policy of the Tenth Plenum

A lengthy *People's Daily* article by Liu Jih-hsin a few months after the Tenth Plenum was a clear exposition of the new policy.[6] The immediate

4. Wang Kuang-wei, "Actively and Steadily Carry out Technical Transformation of Agriculture," *Ching-chi Yen-chiu,* no. 3 (March 17), 1963 (SCMM 361, p. 34). There is, of course, a danger of overdramatizing the sharpness of the break in policy in 1962. There had been important efforts to change agricultural techniques before 1962. And there were important campaigns to reform social relations in the countryside after 1962. The Tenth Plenum also admonished "never forget class struggle," and the Socialist Education Movement came soon after. There was not a total change in policy from social reform to technical transformation in 1962. There was, however, a definite shift in *main emphasis.*

5. Cho T'ien-fu, "On the Question of Intensive Operation of Agriculture in China," *Hung Ch'i,* No. 2–3 (Feb. 4), 1964 (SCMM 407, pp. 14–27).

6. Liu Jih-hsin, "Exploration of a Few Problems Concerning Mechanization of Our Agriculture," *Jen-min Jih-pao,* June 20, 1963 (SCMP 3021, p. 1).

goals were to (1) improve yields, (2) guarantee production of commercial crops, and (3) strengthen management of modern techniques. The previous emphasis on strengthening the collective economy and transforming the rural social and employment structure was subdued.

Liu stressed the need to raise yields in a densely populated country such as China. He also pointed out the complexity and variety of Chinese agriculture. He suggested that agricultural chemicals—fertilizers and insecticides—would be most helpful to raise yields. He wrote, "in the immediate future at least, we should place the universal application of chemicals in a more prominent position than mechanization."

Liu urged that modern inputs be used to assure commodity grain and industrial raw materials needed for industrialization: "First of all . . . we should do our best to increase the production of commercial grain and industrial raw materials so as to satisfy the requirements for the continuous development of the national economy."[7] This would require concentration of modern inputs in specific areas:

> Technical transformation of agriculture and the placement of modern technical equipment and major material resources should not be carried out in such an excessively scattered manner as to seem blooming everywhere. Instead, key points should be decided upon. There should be concentration of forces to fight battles of annihilation, basically winning one battle before waging another.[8]

Specifically Liu recommended: (1) mechanization of cultivation and transport in the Northeast and North, (2) mechanization of irrigation and drainage in the South, (3) supply of chemicals to important regions with concentration of commercial crops, such as cotton, oil, tobacco, sugar, and hemp, and (4) supply of electricity and chemicals to suburban areas.

In accordance with the policy of concentrating inputs geographically where they would be most profitable, a conference on agricultural science and technology met in the spring of 1963 and selected ten major areas to serve as experimental and demonstration sites.[9] These regions were, by and large, the same places that had previously been receiving special attention:

1. The Peking area
2. Northeast China (not more specifically identified)

7. *Ibid.*, p. 11. 8. *Ibid.*

9. " 'Demonstration Farms' Are Main Centers through Which Agricultural Science May Serve Production," *Jen-min Jih-pao* editorial. Oct. 25, 1964 (SCMP 3338, p. 14). The source said ten areas were selected, but listed only seven. It is not known whether there were three more regions, or whether some of the regions mentioned above included more than one demonstration area.

3. The Szechwan basin
4. The Taihu Lake area near Shanghai
5. The Pearl River Delta near Canton
6. Hainan Island
7. Northwest China

Another aspect of the Tenth Plenum policy was a stronger, more direct role for the state. In early 1963, extension stations were re-established, conferences were held on various crops and techniques (such as crop protection), and education and experimentation improved. Also the government took more direct control over agricultural machinery. Liu Jih-hsin thought that large machines, such as tractors, should be owned and managed by state-operated machinery stations or state farms. However, he thought that much smaller equipment—such as 10-h.p. machines—should be owned and operated by production teams.

These various programs were carried out with remarkable speed and effectiveness. By 1965, about 6 million hectares of high- and stable-yield areas had been established, mostly in the Pearl River Delta, lower Yangtze Valley, Peking region, Tungting Lake area, and Szechuan. These regions developed yields comparable to the most modern agriculture in the world. High-yielding dwarf varieties of rice and wheat were developed (incorporating previous resarch), chemical fertilizer was made available, and water control was improved.[10]

These regions could also change their cropping patterns. A detailed report from communes in the Peking suburbs showed how electrified irrigation and application of chemical fertilizers permitted the adoption of improved, more intense crop rotation patterns and the planting of paddy rice. Yields, return on investment, and labor productivity all went up.[11]

While agriculture was modernizing very rapidly in some regions through the application of seed and fertilizer technology, agricultural mechanization was, to some extent, marking time. Previously "the fundamental way out," it was downgraded to one element of the "eight-character constitution." (The "eight-character constitution" was Mao's summary in the late 1950's of the components of agrotechnical transformation: 1, land construction and rearrangement; 2, water control and irri-

10. Benedict Stavis, *Making Green Revolution* (Ithaca: Cornell Rural Development Committee, 1974), chap. 1.

11. Liu Chih-cheng, Ho Kuei-t'ing, and Hsu Hsing, "The Relations between the Four Transformations and Economic Effects," *Ching-chi Yen-chiu*, no. 2, 1964 (SCMM 424, pp. 1–15).

gation; 3, fertilizer; 4, improved varieties of seeds; 5, pest control; 6, improved field mangaement; 7, better farm tools; and 8, suitable spacing of plants.) The time span for mechanization of agriculture was expanded from ten to twenty-five years. A *People's Daily* editorial coming at the close of the February-March 1963 conference described the new plans. "The Party and government have laid down the plan for the technical transformation of our country's agriculture, namely mechanization, electrification, adequate irrigation and drainage, and an ample supply of chemicals for farming, within twenty to twenty-five years."[12]

Since broad mechanization was not imminent, it was recognized that animals would remain the major source of energy for agriculture. Animal husbandry was to be given new emphasis, according to a *People's Daily* editorial published during the conference: "For a relatively long period to come it will be necessary to pay special attention to the development of large livestock, because large livestock is still the *major motive power* for agricultural production in China and the major source of the fertilizer needed in agricultural production."[13] The new policy assumed that mechanization would remain restricted geographically as it had been during the previous few years. Hsiang Nan, one of China's top experts on agricultural mechanization, explained this program:

> The practice of the past few years has proved that according to the procedure for agricultural mechanization, priority should be given: to areas producing staple maketable grain and economic crops, to be followed by areas producing other crops; to units belonging to the system of ownership by the whole people (state farms) to be followed by units of collective ownership; and to methods of cultivation, drainage, irrigation and transportation to be extended step by step to other fields in an over-all manner. Specifically speaking priority should be given to solving problems of mechanizing cultivation in the Northeast and North China, of mechanizing drainage and irrigation in the watered paddy fields of the South, of mechanizing well irrigation in the key pastoral plains of Hopei, Shantung, and Honan, by providing light means of transport suited to the varying topographical and road conditions of the area.[14]

12. "Raise Agricultural Science to a New Level," *Jen-min Jih-pao* editorial, April 6, 1963 (SCMP 2964, p. 6).

13. "Actively Organize and Develop Diversified Agricultural Operations," *Jen-min Jih-pao* editorial Feb. 25, 1963 (SCMP 2933, p. 16). Emphasis added.

14. Hsiang Nan, "Certain Problems of Agricultural Mechanization," *Jen-min Jih-pao,* Dec. 22, 1962 (SCMP 2900, p. 5). Hsiang Nan's several articles outline agricultural mechanization policy throughout the 1960's. In 1976 he was vice president of the China Agricultural Machinery Society and chairman of a delegation of Chinese agricultural machinery specialists which toured the U.S. in September 1976.

To assure full use of available machines, Hsiang Nan urged that investments go to factories making spare parts and implements, as well as to repair shops. He suggested the following proportions for investment:[15]

Tractor factories	100
Implement factories	100
Spare-parts factories	100
Repair shops	150

Of the total budget for agricultural mechanization, only about 22 per cent would go to tractor factories.

Hsiang also recommended many steps to assure the more efficient use of equipment. New designs were still needed for machinery that could operate efficiently under China's particular conditions. Standardization of designs was necessary to reduce the problems of supplying parts to over one hundred types of tractors and internal combustion engines. Quality of equipment had to be improved. Moreoever, management needed to be tightened up. The MTS system was encouraged, and a system by which special personnel would assume full responsibility for each machine was urged. Improved economic accounting was recommended. Closer relations between tractor factories and users would reduce distribution costs, would assure that appropriate machinery was purchased for each locality, and would permit servicing of machines by factory-trained personnel. Finally, Hsiang urged that research into agricultural machinery concentrate on a few pressing problems and that attention be given to the introduction of foreign machinery.

Because agricultural mechanization would require at least twenty-five years, Hsiang emphasized the importance of intermediate steps in the mechanization process—principally semimechanization. He argued that "during the long period of transforming old-type and semi-mechanized agricultural implements into fully-mechanized ones, it is necessary to combine mechanization with semi-mechanization, with emphasis on the use of semi-mechanized and improved implements."[16] This meant, he noted, increasing—not decreasing—reliance on draft animal power during the coming decades. He observed that in the U.S., the USSR, and Japan, animal power actually increased while machinery was being introduced.

It is interesting to note that in his discussion of semimechanization,

15. *Ibid.*, p. 9. 16. *Ibid.*, p. 6.

Hsiang made explicit reference to Mao Tse-tung's interest in tool reform. He recalled Mao's directive of April 1959 concerning the establishment of local tool research institutes and workshops, and the tool-innovation campaign of 1958. Of the latter Hsiang wrote:

> The fervor for reform of implements, which swept the whole country a few years ago, was a roaring mass movement for this reform. When in progress, this movement achieved practical results in many localities and played a great role in increasing production. But there were cases of failure, too. The reason was that the people in one place had followed Comrade Mao's direction but those in another had acted in contravention of it. There was certainly nothing wrong with the movement itself.[17]

The Chinese were very much aware that major questions of social relations were involved in this shift of emphasis from mechanization to chemicals. Liang Hsiu-feng, in an article in *Economic Research* in September 1963, explained:

> Some comrades are of the opinion that since the central aim of agro-technical reform is an increase in farm output and since chemical fertilizers can increase the output with the most striking results, introduction of chemical products should be the center of agro-technical reform. Others take the view that water conservancy "is the life-line of agriculture" and should be the center of agro-technical reform. Still others regard mechanization and electrification as the center of agro-technical reform because mechanization and electrification will not only raise the productivity of agricultural labor with the most notable results but also serve as an important material and technical basis for *changing the rural outlook and consolidating the collective economy of people's commune*. Finally, there are comrades who take the view that mechanization, electrification, water conservancy and introduction of chemical products are of equal importance and are all centers of agro-technical reform.[18]

Mao Tse-tung's position on the choice between machines and chemicals is not clear. He clearly attached great importance to machinery to consolidate the collective economy, to strengthen the worker-peasant alli-

17. *Ibid.*, p. 7. This quotation is another bit of evidence that Mao personally led the tool-reform campaigns of 1958–59 and probably continued to insist on semimechanization during the early 1960's. The admission that failures existed and the charge that they resulted from contravention of Mao's directives is noteworthy. It suggests that Mao may have been disappointed with certain aspects of the Great Leap—perhaps the uniformity of response when his directives had specifically called for careful local testing of techniques. (Others in the Chinese leadership undoubtedly drew different lessons from problems of the Great Leap—probably to Mao's dismay.)

18. Liang Hsiu-feng, "Preliminary Enquiry into the Center, Step and Keypoint of Agro-technical Reform in Our Country," *Ching-chi Yen-chiu*, no. 9 (Sept. 17), 1963 (SCMM 389, p. 36). Emphasis added.

ance, and to transform the nature of rural society. His views of agricultural chemicals, however, seem a bit obscure. As early as January 1957, Mao had enthusiastically (and probably prematurely) announced: "Every province has established its own chemical fertilizer plant."[19] Later, in his letter to cadres at provincial, district, county, commune, brigade, and the team levels of April 29, 1959, Mao mentioned: "It is a matter of great importance to increase chemical fertilizer production year by year."[20]

In the early 1960's, however, just about the time of this debate over the relative importance of machinery and agricultural chemicals, Mao downgraded the importance of chemical fertilizer. He did so not for broad political reasons, but for surprisingly narrow reasons: he feared large supplies of chemical fertilizer would reduce incentives to raise pigs, and he felt that fertilizer derived from hog manure was necessary for maintaining the soil. Mao noted:

> In agriculture we at present do not advocate wide use of chemicals. Firstly, because for years we still will not be able to produce a lot of chemical fertilizers. We already have some chemical fertilizers, but we can only concentrate on using them on industrial crops. Secondly, because if we advocate wide use of chemicals, everybody will set their eyes on this and will not pay attention to raising pigs. We must also have inorganic fertilizer, but if we depend only on it and do not combine it with organic fertilizer, it will harden the soil.[21]

Mao's enthusiasm for pigs had been relayed in a comment he made in October 1959, stressing the role hogs play in recycling.

> A hog is a small scale, organic chemical fertilizer factory. Furthermore, a hog has meat. It has bristles, skin, and bones, and it has viscera which can be used as raw materials for manufacturing medicine. So why don't we take advantage of it? Fertilizer is food for plants, plants are food for animals, and animals, in turn, are food for humankind. Therefore, from this it can be seen that it is definitely within reason to raise hogs and other stock animals on a big scale.[22]

Policy after the Tenth Plenum

Red Guards have charged that "capitalist roaders" undermined agricultural mechanization in the years after the Tenth Plenum. This charge is

19. Mao's Interjections at a Conference of Provincial and Municipal Committee Secretaries, Jan. 1957, *Miscellany,* p. 51.

20. Mao's Intraparty Correspondence, *ibid.,* p. 171. This letter is quoted in Chapter 7, p. 129 above.

21. Reading Notes on the Soviet Union's *Political Economics, ibid.,* p. 288.

22. Mao's Intraparty Correspondence, Oct. 11, 1959, *ibid.,* p. 174.

misleading and shows no understanding for the subtle but important advances in mechanization programs during this period. Agricultural mechanization policy certainly changed, but it was not undermined. The new policy emphasized central planning and control of farm machinery with regard to investment, research, management, and distribution. It did not, however, sharply increase supplies of machines.

Relation to Industrial Policy

The new policy did undermine the vision put forward at Chengtu, in which agricultural mechanization was to become a cornerstone of rural social development and rural industry. In contrast, policy after the Tenth Plenum emphasized specialized, highly efficient factories responsible directly to the central ministries. This type of industrial development is reflected in the investment patterns after 1962. A report in 1963 indicated that much of the investment was going to develop production of spare parts: "In this year's capital construction for agricultural machinery industry, factories producing spare parts occupy a portion which is the largest of recent years. Also, on the average, funds invested in several principal spare part factories increased by one to over three times compared with the past few years."[23] Investments were announced in July 1963 for six major specialized factories that would make accessories (fuel pumps, pistons, rings, generators, regulators) for internal combustion engines.[24]

Development of these specialized parts factories is significant because they represented an innovation in overall industrial structure. Previously, each factory tended to make within its walls all the parts for a tractor. (The prototype for tractor manufacture in this style was the Loyang tractor factory, set up by the Soviet Union.) These new specialized factories were undoubtedly related to an observation Hsiang Nan had made in his article of December 1962 concerning industrial structure in advanced capitalist countries:

In order to speed up the development of the agricultural-machinery industry, we should give a thought to the experience gained in various countries of the world

23. Chun Wen, "New Developments in the Production of Agricultural Machinery," *Kung-jen Jih-pao,* Sept. 24, 1963 (SCMP 3089, p. 16).

24. "China Doubles Capital Investment in Farm Machinery Accessories Plants," NCNA Peking, July 31, 1963 (SCMP 3033, p. 13); "China Speeds up Construction of Six Major Farm Machinery Accessories Plants," *Jen-min Jih-pao,* Aug. 1, 1963 (SCMP 3041, p. 14); "China's First Factory Producing Electrical Machinery for Tractors," NCNA Chengchow, Sept. 28, 1964 (SCMP 3309, p. 19); "China's First Tractor Accessories Plant," NCNA Hofei, Jan. 16, 1965 (SCMP 3381, p. 20).

today that the degree of specialization is getting higher and higher and the scope of cooperation, wider and wider. Specialization and cooperation have the advantages of achieving a large output, turning out products of high quality, lowering the cost of production and solving technical problems with ease. Under ordinary conditions, the machine parts (such as valves, cylinder bodies and pistons) made by specialized factories are approximately 50 percent cheaper and require 50 percent less labor than those made by all-purpose factories. The common characteristics of the agricultural machinery industry in the United States, Britain, West Germany, and Japan are the high degree of specialization and the large numbers of small factories which sell exclusively to a few large factories. Their tractor factories are, in reality, assembling plants.[25]

These specialized factories were operated by the central government. While emphasis was on spare-parts factories, investments were also made in tractor factories that would supply medium-sized and small tractors. A factory in Shenyang produced its first 28-h.p. tractors in April 1963, and experimental production of a 7-h.p. walking tractor began in Shanghai in early 1964.[26]

During this period, efforts were made to improve the quality of agricultural machinery. The Ministry of Agricultural Machinery held a national conference of directors, secretaries of Party committees, and chief engineers of agricultural machine works in May 1963 to emphasize the importance of quality in the manufacture of machinery.[27]

There are indications that this shift in the pattern of industrial development coincided with the question of the relationship between coastal and inland industry.[28] One Red Guard critique mysteriously suggested that location and scale of agricultural machinery factories constituted one of the features of policy after 1962 which distressed Mao:

What was called the "Battle of XX" was a salient "model" engineered by this counter-revolutionary revisionist line of theirs. The melee that lasted X years and wasted the sum of XXX thousand yuan disrupted Chairman Mao's strategic plan concerning the relationship between coastal industry and inland industry, and

25. Hsiang, "Certain Problems of Agricultural Mechanization," SCMP 2900, p. 9.

26. P. H. M. Jones, "Machines on the Farm," *Far Eastern Economic Review* 45 (1964), 479.

27. "Ministry of Agricultural Machinery Requests Industrial Enterprises to Continue to Improve Quality of Agricultural Machines," *Jen-min Jih-pao,* May 9, 1963 (SCMP 2987, p. 4).

28. Mao had discussed the regional pattern of investment in earlier years and had warned against putting too many investments inland. "On the Ten Great Relationships," April 25, 1956; available in Stuart Schram, ed., *Chairman Mao Talks to the People* (New York: Pantheon, 1974), pp. 65–67.

edged out the construction of large numbers of medium and small enterprises, principally local ones.

As a result . . . the construction of China's agricultural machinery industry greatly lags behind the need for the development of agricultural mechanization. Not a single area has established a relatively complete and independent system.[29]

Research and Development

A similar policy of specialization and centralization evolved in agricultural mechanization research. Most of the roughly 600 tool-research institutes established before 1961 were closed. By 1965, there were only 92 left, the others having been dissolved in the general financial squeeze.[30] A criticism made in early 1962 of the Kiangsi Province Agricultural Machinery Research Institute cited as a principal weakness in agricultural machinery research "too many topics and too few practical results. . . . The major defect of the research program is the insufficient attention to reality. There are not enough surveys concerning the actual conditions and there is too great a desire to copy foreign products and to achieve a great deal within a short time."[31]

At the same time, however, central technical capabilities to coordinate research were strengthened. In May 1962 there were five research institutes directly supervised by the Ministry of Agricultural Machinery; over 10,000 researchers were working exclusively on agricultural mechanization.

Soon afterward, probably in 1963, researchers established eight farming regions, each with its own machinery requirements.[32]

1. Northeast—ridge farming
2. North China Plain and Huai River Plain—flat, dry-land farming
3. South China—multiple-crop paddy fields, clay soil
4. Southwest—paddy and dry fields
5. Northwest—irrigated by rivers, snow, and glaciers
6. Loess highlands—soil erosion problems

29. Eighth Ministry of Machine Building, Research Office of the United Headquarters, Support the Peasants and Soldiers Fighting Detachment, "China's Agricultural Machine Industry Must Take the Road Indicated by Chairman Mao," *Nung-yeh Chi-hsieh Chi-shu,* no. 9, 1968 (SCMM 644, p. 26).

30. "History of Struggle," SCMM 633, p. 21.

31. Li Te-yu, "Scientific Research Should Start from Reality," *Chung-kuo Nung-yeh Chi-hsieh,* no. 5 (May), 1962, pp. 21–22 (JPRS 43,912, p. 35).

32. Li Ch'ao, "Opinions Concerning Research Work in Agricultural Mechanization," *Chung-kuo Nung-yeh Chi-hsieh,* no. 5 (May), 1962, pp. 5–8 (JPRS 43,912, p. 31); Ching Tan, "Machinery Especially for Chinese Farms," *China Reconstructs,* Aug. 1964, p. 6.

7. Inner Mongolia and far west—pasturelands
8. Tropical crop areas

There were plans for major testing facilities in each region.

Farm machinery research institutes and centers were established in every province and autonomous region, and in many large cities and counties.[33] These were specifically for research in agricultural mechanization, and were in addition to general agricultural research institutes. Provincial agricultural mechanization research institutes are listed (incompletely) in Table 8.1.

To further coordinate high-quality research, the Chinese Agricultural Machinery Society was formally established on March 3, 1963.[34] (A preparatory committee had been formed in 1956.)[35] Its top leadership included men from government and academic posts. The society published the *Journal of Agricultural Machinery,*[36] and held its first annual convention in Nanking in January 1964, at which time various machines and tools for paddy cultivation were examined and reviewed.[37] Other conferences were held in November 1962 and in August 1964.[38] The society sponsored the design and trial manufacture of new farm implements. By January 1965 work had been done on equipment for sowing, interrow cultivation, weeding, harvesting, husking, and shelling.[39]

During 1963 and 1964, important progress was made in farm-tool research and development. Many tools, especially tractor-drawn implements, were designed and produced to meet farming requirements in North China. Among these were: large four- and seven-blade plows, forty-eight-row seeders, equipment to level land and to combine spreading fertilizer with sowing, and special tools suited for North China's peculiar system of ridge cultivation.[40]

33. Ching, *ibid.;* "China Uses More Tractors and Agricultural Machinery," NCNA Peking, Oct. 20, 1964 (SCMP 3324, p. 18).

34. "Chinese Agricultural Machinery Society Inaugurated in Peking," NCNA Peking, March 5, 1963 (SCMP 2936, p. 10).

35. *Ibid.* 36. *Ibid.*

37. "First Annual Convention of Chinese Agricultural Machinery Society," NCNA Nanking, Jan. 2, 1964 (SCMP 3133, p. 20).

38. Sun Wei, "The Annual Conference of the National Society of Agricultural Machinery Is in Preparation," *Chung-kuo Nung-yeh Chi-hsieh,* no. 9 (Sept.), 1962, p. 8 (JPRS 43,912, p. 61); "National Conference on Mechanics of Agricultural Machinery," NCNA Tsingtao, Aug. 20, 1964 (SCMP 3286, p. 17).

39. "New Maize Machinery for China's Agriculture, NCNA Peking, Jan. 7, 1965.

40. "Northeast China Province Becomes Research Center for Agricultural Mechanization," NCNA Harbin, May 16, 1963 (SCMP 2984, p. 19); "Tractor-Drawn Machines for Northeast China Fields," NCNA Peking, May 29, 1963 (SCMP 2992, p. 17); Chun, "New

Table 8.1. Provincial agricultural machinery research institutes (incomplete), 1963–65

Province	Working on
Heilungkiang[a]	
Liaoning[b]	
Inner Mongolia[c]	Stock-breeding equipment
Shantung[d]	Tractor-drawn plows
Kiangsu[e] (branch of Chinese Academy of Agricultural Sciences at Nanking)	
Chekiang[f]	Plow for paddy
Kwangtung[g]	Rice combine harvester
Kweichow[h]	Pumps, plows, reapers, husker
Szechwan[i]	Pump
Hopei[j]	
Hunan[k]	Paddy plow

Sources:

a. "Northeast China Province Becomes Research Center for Agricultural Mechanization," NCNA Harbin, May 16, 1963 (SCMP 2984, p. 19).

b. "Northeast China Peasants and Craftsmen Devise Farm Machines," NCNA Shenyang, Oct. 30, 1965 (SCMP 3571, p. 21).

c. "Inner Mongolia Begins to Mechanize Stockbreeding," NCNA Huhehot, Dec. 30, 1963 (SCMP 3133, pp. 21–22).

d. "East China Province Mechanizing Farming," NCNA Tsinan, Oct. 22, 1963 (SCMP 3088, p. 10).

e. "New Machines for South China Paddy Fields," NCNA Nanking, Jan. 10, 1965 (SCMP 3377, p. 18).

f. "Five New Types of Tractor-Drawn Plows in China," NCNA Peking, May 28, 1965 (SCMP 3470, p. 17).

g. "China Makes Rice Combine-Harvester," NCNA Canton, Sept. 4, 1965 (SCMP 3534, p. 21); Chou Chin-ken, Kwangtung Provincial Agricultural Machinery Research Institute, "Rice Paddy Farming and Farming Machines," *Chung-kuo Nung-yeh Chi-hsieh,* no. 1 (Jan.), 1965, pp. 22–24 (JPRS 42,484, p. 6); Yeh Ping-ch'iang, "Sugar Cane Defoliator," *Nung-yeh Chi-hsieh Chi-shu,* no. 11 (Nov.), 1963, p. 15 (JPRS 44,251, p. 29).

h. "Specialized Farm Machinery and Tools for China's Mountainous Regions," NCNA Kweiyang, Nov. 19, 1965 (SCMP 3584, p. 12).

i. "Chinese Factories Make Big Efforts to Improve Service to Rural Areas," NCNA Peking, Dec. 4, 1965 (SCMP 3593, p. 19).

j. Hopei Provincial Institute of Agricultural Machinery, "Organized Cooperation Promotes the Development of Agricultural Machinery Research Work," *Chung-kuo Nung-yeh Chi-hsieh,* no. 9 (Sept.), 1962, pp. 16–17 (JPRS 43,912, p. 65).

k. Office of the Hunan Institute of Agricultural Machinery, "The Growth of the 55 Type Improved Paddy Plow," *Chung-kuo Nung-yeh Chi-hsieh,* no. 6 (June), 1962, pp. 19–20 (JPRS 43,912, p. 51).

Progress was also made in designing and trial-producing equipment for South China. By 1963 a small 10-h.p. diesel engine had been designed and mass-produced in Hupeh and Nanchang; the following year it was reported being manufactured in Szechwan.[41] The engine provided power for small pumps, food-processing equipment, and electric generators. At the same time, work was done on small tractors. By January 1963 a prototype for the 7-h.p. "Kung-Nung" (Worker-Peasant) walking tractor had been developed at Shanghai, and by 1965 it was being made at Wuhan and Shenyang.[42] Implements for this small tractor were manufactured by the Shanghai Tractor Plant and elsewhere.[43]

Research continued on the rice transplanter, prematurely popularized in 1960. By the end of 1962 five models were being manufactured and tested, but the machines were not yet more efficient than hand transplanting.[44] Continued research was reported in 1963, and review in 1965 indicated that, while progress had been made, a satisfactory transplanter was not yet available.[45]

Developments in the Production of Agricultural Machinery," SCMP 3089, p. 16; "Peking Industry Turns Out Fertilizer, Farm Equipment for Rural Communes," NCNA Peking, Aug. 18, 1964 (SCMP 3284, p. 14); "Combined Seeder-Manurer Introduced on Northeast China State Farm," NCNA Harbin, May 22, 1965 (SCMP 3465, p. 25); "New Types of Farming Machines Manufactured in China," NCNA Peking, Jan. 4, 1964 (SCMP 3134, p. 17).

41. "Machine Building Workers in Yangtze Port City Aid Agriculture," NCNA Wuhan, Feb. 25, 1963 (SCMP 2929, p. 14); "Industrial Work Gradually Brought in Line with Agriculture as the Foundation," *Kuang-ming Jih-pao,* Jan. 21, 1963 (SCMP 2925, p. 1); "Southwest China Province Steps Up Production of Farm Machines," NCNA Chengtu, May 2, 1965 (SCMP 3451, p. 21).

42. "Shanghai Engineering Helps Agriculture," NCNA Shanghai, Jan. 12, 1963 (SCMP 2899, p. 13); "Central China City Produces Walking Tractors," NCNA Wuhan, Feb. 2, 1965 (SCMP 3392, p. 14); "More Farm Machinery Produced in Northeast China Province," NCNA Shenyang, April 6, 1965 (SCMP 3435, p. 23). Output figures of the Wuhan tractor factory for this tractor are supplied by Bruce McFarlane, who visited the factory in 1968: In 1964, the factory produced 50 tractors; in 1965, the factory produced 500; in 1966, 1,500; and in 1967, 3,000 (unpublished diary of Bruce McFarlane, April 26, 1968).

43. Ting Shao-ch'eng, "Some Accessory Farm Implements for Manual Tractors," *Chung-kuo Nung-yeh Chi-hsieh,* no. 8 (Aug.), 1965, cover 2 (JPRS 42,545, p. 76); "New Types of Farming Machines Manufactured in China," NCNA Peking, Jan. 4, 1964 (SCMP 3134, p. 17); "Turbine Pumps for South China Autonomous Region," NCNA Nanking, March 9, 1964 (SCMP 3177, p. 14); "New Machines for South China Paddy Fields," NCNA Nanking, Jan. 10, 1965 (SCMP 3377, p. 18).

44. Lin T'i-ch'iang, "The Current Condition of Development of Rice Transplanting Machines," *Chung-kuo Nung-yeh Chi-hsieh,* no. 9 (Sept.), 1962 (JPRS 43,912, pp. 63–64).

45. Seedling Transplanting Machinery Group, Department of Agricultural Machinery,

Research was also conducted in plant studies. The Chinese Crops Association held a conference during 1963 to discuss crop-rotation systems, soil fertility, and types of tractors suitable for various patterns of cultivation.[46]

Importing Technology

In conjunction with research work, the Chinese specialists were, at this time, very much interested in the development of agricultural machinery in other countries. In fact, Red Guards charged that the "capitalist roaders" were imitating the equipment in foreign countries. Specifically, the "capitalist roaders" are said to have issued a "revisionist" instruction: "The Northeast should learn from the Soviet Union, North China from West Germany, and South China from Japan."[47]

While this instruction has not been located, it is clear that such sentiments existed in China in 1963. An article in *Economic Research* about electrification said:

> The Soviet and American experiences pertaining to the modernization of agricultural techniques are applicable principally to a few zones, including Heilungkiang. . . . In East China, Central South China and Southwest China where grain yield is high, and in the plains of North China . . . where a number of principal conditions for agricultural production resemble those of Japan, the Japanese experiences are of reference value.[48]

Kiangsi Provincial Institute of Agricultural Sciences, "The Process of Improving the Kiangsi-59 Type Transplanting Machine," *Chung-kuo Nung-yeh Chi-hsieh,* no. 8 (Aug.), 1963, pp. 15–16 (JPRS 43,905, p. 41); Planting Machine Team of Agricultural Machine Examination Station of Kiangsi Provincial Agricultural Department, "Experiments on the Use of Rice Seedling Planting Machines for Transplanting," *Chung-kuo Nung-yeh Chi-hsieh,* no. 4 (April), 1965, pp. 27–29 (JPRS 42,484, p. 19).

46. "The First Conference on Mechanized Cultivation Held by the Chinese Crops Association," *Chung-kuo Nung-yeh K'o-hsüeh,* no. 8 (Aug.), 1963, p. 54 (JPRS 42,524, p. 16).

47. China Scientific Research Institute in Agricultural Mechanization, Criticism and Transformation Group of the Revolutionary Committee, "Seizing Back Power from the Bourgeoisie in the Scientific and Technical Field," *Nung-yeh Chi-hsieh Chi-shu,* no. 9 (Sept.), 1968, supplement (SCMM 632 p. 10). A scathing attack on "capitalist roaders" (particularly Liu Shao-ch'i) for wanting to import technology in the electronic industry is available in Chao I-p'ing, "Down with Foreign-Slave Philosophy; Take the Road of Self-Reliance," *Kuang-ming Jih-pao,* March 22, 1969 (SCMP 4395, pp. 4–5), and Lao T'ung-ping, "Thoroughly Bury Foreign-Slave Philosophy," *Kuang-ming Jih-pao,* March 22, 1969 (SCMP 4395, pp. 6–7).

48. Tso Hu, "Some Problems of Agricultural Electrification," *Ching-chi Yen-chiu,* no. 3 (March 17), 1963 (SCMM 360, p. 2).

It is also clear from the agricultural machinery journals that Chinese followed closely developments abroad. Articles appeared about machinery in the Soviet Union, the exhibits at the Second International Agricultural Machinery Fair in Paris in 1963, Italian machinery for rice fields, and machinery in Japan, Korea, England, and France.[49]

Red Guards charged that foreign machines were imported and copied. Capitalist roaders supposedly proposed that "designing must be faithful to the sample machines," and "sample machines should be copied down to every detail." Moreover, the Soviet Union's encyclopedia and *Handbook on Designing of Tractors and Trucks* were reportedly used as textbooks and praised as "two god-sent books."[50] Red Guards also charged that "the agents of China's Khrushchev in the farm machinery system purchased from France and Japan, the two imperialist countries, patents for liquid pressure equipment and small-sized gasoline machines and wasted a large amount of foreign exchange in buying a large quantity of general equipment and blueprints which were of little use."[51]

These charges have a certain element of truth but are not entirely correct. Of course the Number 1 Tractor Factory, at Loyang, had been designed and installed by Soviet technicians, so almost all of its technology came directly from the Soviet Union.[52] Before 1963, Chinese imports of agricultural machinery from the West were virtually nonexistent. Then, in 1963 and 1964, somewhat over $1 million worth of machinery was imported from Japan. The wide variety and small quantities might in-

49. Tseng Chung-ch'ih, "Foreign Agricultural Machinery Technology," *Chung-kuo Nung-yeh Chi-hsieh,* no. 4 (April), 1962, p. 30 (JPRS 43,912, p. 27); Wang Wan-chun, "Current Trends in the Development of Farm Implements in Western and Northern Europe," *Nung-yeh Chi-hsieh Chi-shu,* no. 11 (Nov.), 1963, pp. 2–4 (JPRS 44,251, p. 25); Hu Chung, "Foreign Machinery," *ibid.,* pp. 12–13 (JPRS 44,251, p. 27); Ts'ai Shao-chen, "Several Water Pumping Machines of Japan," *Chung-kuo Nung-yeh Chi-hsieh,* no. 5 (May), 1964, p. 31 (JPRS 43,905, p. 57); Cheng Ch'i-chen and Shih Hsiang, "The General Condition of Agricultural Mechanization in Japan," *ibid.,* no. 8 (Aug.), 1963, pp. 31–32 (JPRS 43,905, p. 47); Li Ching-mo, "Production Condition of Agricultural Machinery in Western Europe," *ibid.,* pp. 27–30 (JPRS 43,905, p. 47).

50. "History of Struggle," SCMM 633, p. 25; "China's Agricultural Machine Industry Must Take the Road Indicated by Chairman Mao," SCMM 644, p. 24.

51. "History of Struggle," SCMM 633, p. 26.

52. At the Third Session of the Sino-Soviet Commission for Scientific and Technical Cooperation, meeting in Moscow in December 1955, it was agreed that the Soviet Union would supply China with working drafts for the production of various types of agricultural machinery. M. I. Sladkovskii, *History of Economic Relations between Russia and China* (Moscow, 1957; translation, Jerusalem: Israel Program for Scientific Translations, 1966), p. 270, n. 29; reported by NCNA Moscow, Jan. 4, 1956. (SCMP 1203, p. 24).

dicate that the machines were being imported to be studied and perhaps duplicated.[53]

There were a few other purchases of agricultural machinery, all small. From the United Kingdom, China bought about $100–150 thousand of equipment each year from 1964 to 1967. There were no imports from West Germany (even though Red Guards said it would be a model) until the purchase of $176 thousand of harvesting equipment in 1969.[54] These imports from Japan and the United Kingdom were undoubtedly used as models. Chiang Tan, writing in *China Reconstructs,* pointed out that "the basic approach has been to select traditional implements or foreign models and modify those which give good performance but are not entirely adapted to specific conditions."[55]

As for the specific charge that patents were purchased from France and Japan for liquid-pressure equipment and small-sized gasoline machines, information is scarce. It is known that China obtained a license from a French company for the manufacture of trucks and tractors in 1965.[56] Moreover, it is interesting to note that in January 1966, China announced

53. Chinese imports of agricultural machinery from Japan in 1964 were:

Item	Number	Value (U.S. $ F.O.B.)
Harrows	6	645
Cultivators	40	610
Power tillers	24	25,200
Soil preparers	56	6,200
Parts for above		71,400
Power threshers	17	3,610
Harvesting equipment	53	38,600
Parts for above		536
Two-wheel tractors	5	5,860
Wheel tractors	2	2,300
Track-laying tractors	80	1,113,000
Presses		5,870
Bee and poultry equipment		58,000
Total		$1,315,731

From *Japan Commodity Trade Statistics* (Tokyo: Government of Japan, 1964).

54. *Commodity Trade Statistics*. (Washington: International Monetary Fund, various years).

55. Ching, "Machinery Especially for Chinese Farms," p. 6.

56. Letter to author from M. Lanier, of the French Ministry of Industrial and Scientific Development, July 9, 1971.

production of seven types of small, lightweight gasoline engines in the 1.5–5.5 h.p. range, both air-cooled and water-cooled, both two-cycle and four-cycle.[57] Also, interestingly enough, four models of hydraulic pumps were introduced in 1965.[58]

It appears correct that after the Tenth Plenum China "imported" technology. However, to call this a "slave mentality" and a "go-slow philosophy," as Red Guards did, is inaccurate. Technology was imported precisely to speed up the process of mechanizing agriculture. Moreover, there never was any indication that China was planning to rely on imported equipment or give up its own independent manufacturing or engineering capabilities. On the contrary, these were strengthened during this period.

Despite Red Guard charges, this policy did not violate Mao's views on importing technology. Mao did not want China to lag behind, or blindly imitate: "We should not follow the old road of technical development of other countries and crawl one step after another behind other people."[59] At the same time, Mao insisted that China must not ignore progress in other countries:

> There are two different attitudes toward learning from others. One is the dogmatic attitude of transplanting everything, whether or not it is suited to our conditions. This is no good. The other attitude is to use our heads and learn those things which suit our conditions, that is, to absorb whatever experience is useful to us. That is the attitude we should adopt.[60]

Was Mechanization Undermined?

While these important steps were being taken to rationalize and develop the agricultural machinery system, actually rather little was done in terms of increasing the supply of tractors. (This was not true of mechanized irrigation equipment, which grew dramatically during this period.) After 1960, tractor production dropped somewhat, and then rose only slightly until 1964. Since the Loyang tractor factory alone had an output

57. "Several Types of New, Compact Gasoline Engines," *Nung-yeh Chi-hsieh Chi-shu,* no. 1 (Jan.), 1966, p. 10 (JPRS 34,636, p. 5).

58. "Performance Specifications of Four Types of Hydraulic Pumps," *Chung-kuo Nung-yeh Chi-hsieh,* no. 8 (Aug.), 1965, p. 34 (JPRS 42,545, p. 72).

59. Eighth Ministry of Machine Building, Revolutionary Great Alliance Headquarters and Revolutionary Great Criticism and Repudiation Group of Organizations, "Two Diametrically Opposite Lines in Agricultural Mechanization," *Nung-yeh Chi-hsieh Chi-shu,* no. 9, 1968 (SCMM 633, p. 42).

60. Cited in *ibid.*

of about 15,000 units and started serial production in 1959, the data shown in Table 8.2 suggest that significant new productive facilities were not opened until 1964.[61]

One Red Guard charge suggests that Po I-po resisted efforts to expand the productive capacity of farm machinery industry: "Po I-po still insisted on abolishing seven backbone enterprises in the farm machinery industry and persistently prevented farm machinery departments from setting up centers for forging and casting, assembly and machine repair,

Table 8.2. Production of tractors, 1957–64

Year	Tractors produced
1957	0
1958	957
1959	5,598
1960	24,800
1961	15,200
1962	14,800
1963	17,800
1964	21,900

Source: Chao, *Agricultural Production,* p. 107.

trying in all possible ways to restrict the development of the farm machinery industry."[62] Red Guards also charged that money budgeted by central authorities for agricultural machinery was not spent for these purposes in some provinces: "In Shantung, from 1959 to 1964, some 80 per cent of the funds earmarked for procurement of drainage and irrigation machinery were misappropriated."[63] Szechwan, according to the same source, refused the offer of tractors, and asked instead for investment

61. A listing of 44 factories that manufactured (or had planned to manufacture) tractors, tractor parts, and farm implements before 1965 is available in Chi I'chai, "A Study on the Production and Need of Agricultural Machinery on China Mainland," *Fei-ch'ing Yueh-pao* [Monthly Journal on Bandit Situation] 10:3 (May 31, 1967), 63–72 (JPRS 42,271, pp. 1–21; the listing is at pp. 15–19).

62. "History of Struggle," SCMM 633, p. 19. Of course Po I-po may have have "abolished" some enterprises and prevented expansion of others because he thought they were inefficient, or the final product was not technically suitable.

63. *Ibid.,* p. 21. This corresponds with the charge that T'an Chi-lung, governor and first secretary of Shantung, feared mechanization of irrigation would generate labor surpluses. This charge is quite interesting in view of the indication that Shantung was one of the major recipients of tractors, but no major mechanized irrigation projects have been reported.

funds. There is no independent evidence confirming or denying these charges, but they appear to be internally consistent.

With regard to the availability of small tools, there is a bit of data showing some shortages. A survey taken in Liuyang County, Hunan, by the Supply and Marketing Cooperative in 1965 on tool supply revealed that, on the average, poor and lower-middle peasants owned 8.8 tools, while rich peasants owned 13.6 tools.[64] (It is not clear whether the ownership was individual or collective.) The reporters of the survey considered that the poor and lower-middle peasants required an average of 5.6 more tools. It appears that the number of tools available in the 1960's was not larger than it had been in the 1930's.[65]

While there was no dramatic increase in the supply of agricultural machinery after 1962, it must be remembered that this was a period of severe depression. Industrial production dropped roughly one-third in 1961. Not until 1966 was the 1960 index of total industrial production reached. Thus the fact that agricultural machinery production grew at all indicates that it was receiving special priority.

In fact, there is some evidence that the relative importance of farm machinery compared to other industries grew somewhat during these years. From 1956 to 1966, agricultural machinery grew from 7.0 per cent of the machine-building output to 10.0 per cent. During this time the total machine-building industry grew by more than two and one-half times.[66] In Shanghai, agricultural machinery grew from 5.3 per cent of total output of engineering industries in 1958 to 16.1 per cent in 1962.[67]

More important than gross indications of overall growth was the fact that the agricultural machinery industry improved in quality at this time. It moved toward a rational balance of production of new machines, spare parts, and repair stations. Research improved, and machines and imple-

64. Investigation Team of Liuyang County Supply and Marketing Cooperative, Hunan, and Yang Yüan-hou (*Ta-kung Pao* correspondent), "To Help Poor and Lower-Middle Peasants Solve the Problem of Small Tools Is a Matter of Urgency," *Ta-kung Pao,* March 6, 1975 (SCMP 3422, p. 17).

65. John Buck's survey of five sites in Hunan in 1930 showed that farm families had roughly the same number of tools: small farms averaged 7.0 tools, medium farms averaged 9.4 tools, and large farms averaged 20.4. John Buck, *Land Utilization in China, Statistics* (Chicago: University of Chicago Press, 1937), p. 399.

66. Cheng Chu-yuan, "Growth and Structural Change in the Chinese Machine Building Industry, 1952–66," *China Quarterly,* no. 41 (Jan.–March 1970), 39.

67. Chin Chung-hua, "China's Biggest City Aids Agriculture," *China Reconstructs,* April 1963 (SCMM 361, p. 45).

ments were designed for specific Chinese needs. During this period after the Tenth Plenum the basis was laid for future expansion of agricultural machinery.

Setting Up Trusts to Improve Management

To further coordinate the new programs of research and development, importing technology, manufacturing machinery, and utilizing it efficiently a new managerial system was established after the Tenth Plenum, namely, the system of trusts. Mao strongly opposed this system, probably perceiving it as a major—perhaps irrevocable—step toward revisionism and state capitalism, as practiced in the Soviet Union. However, Mao was loosing his political control over administrative matters and could not block the implementation of this system until he created the extraordinary Cultural Revolution to stop this and other practices he considered revisionist.

Problems of Agricultural Machinery Stations during the Early 1960's

The tractor stations that were re-established in 1961 and 1962 had the same general institutional form as those that existed before 1957. The name was changed from machine tractor station (MTS) to agricultural machinery station (AMS), however, indicating a slightly broader function.[68]

More importantly, it appears that serious efforts were made to establish the AMS on a cost-accounting basis, so that the growing agricultural machinery system would be self-supporting and would not require financial subsidies from the state.[69] With a vastly enlarged system—at the end of 1963 there were 68,040 standard tractors, almost six times the number in 1957—the potential financial losses were much greater, and this was a cause of concern to the Chinese leadership. Hsiang Nan inspected tractor

68. The widening of function is suggested by Kang Chao in *Agricultural Production in Communist China, 1949–1965* (Madison: University of Wisconsin Press, 1970), p. 111. It is also suggested by Audrey Donnithorn in *China's Economic System* (New York: Praeger, 1967), p. 112.

69. In the Soviet Union in 1956–57 there was considerable discussion about putting the MTS system on a cost-accounting basis for similar reasons. The plan was not adopted, and instead the MTS's were dissolved and machinery sold to the kolkhozy. Roy Laird, Darwin Sharp, and Ruth Sturtevant, *The Rise and Fall of the MTS as an Instrument of Soviet Rule* (Lawrence: Governmental Research Center, University of Kansas, 1960), pp. 39–40, 81. Undoubtedly, Chinese administrators were well aware of these discussions in the Soviet Union.

stations in 1963–64 and found several reasons why stations might lose money.[70] One important consideration was whether or not they had the proper mix of large and small tractors and implements for the local geographic and cultivation conditions throughout the year. If the stations were to show a profit these machines and implements had to be kept in good repair so they could be used at the appropriate times.

Staff expense was another factor affecting profitability. When Hsiang toured tractor stations in 1963–64, he observed that

> in some counties a management station with several dozen or up to a hundred tractors has nevertheless established several sets of organs and employed a great number of men. The greater the number of men employed, the greater the procrastination, the higher the costs, and the lower the efficiency. This is also an important reason for the financial losses sustained by agricultural machinery stations in some places.[71]

So excessive was the personnel in some places that mechanization was not reducing the labor requirements of agricultural production but merely changing the requirements from farm laborer to tractor-station worker.

T'an Chen-lin, according to Red Guards, was very concerned about station losses. At a specialized conference on agriculture in 1963 he asked that farm machinery stations and drainage and irrigation stations be given two years in which to turn their losses into profits. "Within two years, they shall not be allowed to incur any losses. Those which continue to lose shall be abolished, while those which operate with profit shall continue to be developed."[72] The Red Guards said he made unauthorized use of the name of the Central Committee when he said, "The Central Committee wants to see no losses but profits."[73] In Kiangsu provincial authorities did try to cut losses: "No machines will be assigned to those who sustain heavy losses. Those who are incompetent must hand over control to other parties. If this still does not work, both the machines

70. Hsiang Nan, "Agricultural Mechanization Can Be Achieved with Good and Fast Results," *Jen-min Jih-pao,* July 6, 1965 (SCMP 3518, p. 7). The full article, from which the *Jen-min Jih-pao* article was excerpted, was published in *Chung-kuo Nung-yen Chi-hsieh,* May 1965, pp. 11–20, and is available in translation at JPRS 33,691.

71. *Ibid.,* SCMP 3518, p. 7.

72. "History of Struggle," SCMM 633, p. 30. Also available in: Eighth Ministry of Machine Building, Agricultural Machinery Management Bureau, Revolutionary Great Alliance Committee, Mass Criticism Unit, "T'an Chen-lin's Crime of Sabotage against Farm Mechanization Must Be Reckoned With," *Nung-yeh Chi-hsieh Chi-shu,* no. 6 (June 8), 1968 (SCMM 624, p. 2).

73. *Ibid.,* SCMM 624, p. 2.

and the personnel must leave.''[74] Likewise in Heilungkiang plans were announced to dissolve stations which did not make a profit.[75]

The mood of the early 1960's was captured in the title of an article in *Chinese Agricultural Machinery* in March 1963: ''P'inglopao Tractor Station Makes Profit Every Year.'' It was subsequently reported that tractor stations in Liaoning launched a ''Learning from and Overtaking P'inglopao'' campaign.[76]

How the losses were subsidized is not known for certain. The interest of T'an Chen-lin, Politbureau specialist on rural questions, combined with the fact that the stations were operated as agents of the central government (operating through the Ministry of Agricultural Machinery), suggests that losses were subsidized by the central government. It is also possible that provincial and even county governments met part of the losses. There may, however, have been regional differences and experimentation. Sometime in early 1963, the Kirin Provincial Agricultural Administration implemented the policy of ''charging the county with full financial responsibility'' for the operation of tractor stations. While this may mean only that the county audited the reports of the stations, it may also mean that the county had to meet the stations' losses.[77]

Some stations may have been closed after T'an Chen-lin's threat. The growth in the number of tractor stations from 1963 (1,482) to 1964 (1,488) was virtually nil and efficiency improved. The farm machinery stations in 1964 did 42 per cent more work than in 1963, and each standard tractor increased its work output by 18 per cent. At the same time, the cost for cultivating each standard mou dropped by 14 per cent.[78] In 1965, however, the number of new stations increased sharply (from 1,488 to 2,263) and with this increase came the possibility of increased losses.

74. Red Workers' Combat Detachment Defending the Thought of Mao Tse-tung, Agricultural Machinery Corporation, Hsinyi County, Kiangsu, ''Refute Material Incentive and Render It Repugnant,'' *Nung-yeh Chi-hsieh Chi-shu*, no. 4 (July 8), 1967 (SCMM 605, p. 27).

75. Take-over Committee of the Agricultural Machine Station, Nunchiang County, Heilungkiang, ''Thoroughly Eliminate the Pernicious Influence of Material Incentive,'' *ibid.* (SCMM 605, p. 30).

76. ''Further Improving the Care and Use of 'Iron Cows,' '' *Chung-kuo Nung-yeh Chi-hsieh*, no. 6 (June 10), 1963 (SCMM 374, p. 11).

77. *Ibid.*

78. ''Opening up the Road of Agricultural Mechanization in Our Country,'' *Jen-min Jih-pao*, Aug. 31, 1965 (SCMP 3543, p. 7).

Another dimension in the change from MTS's to AMS's was increased use of material incentives for station personnel. In 1960, T'an Chen-lin recommended that rewards be given for exceeding the acreage targets per tractor.[79] During the early 1960's workers in the AMS's could supplement their salaries with bonuses, which were given for savings in fuel and maintenance and for exceeding quotas. The rewards were supposed to be one-third or one-half of the savings.[80]

T'an considered fuel economy so important that in 1960 he recommended: "When the consumption of gasoline is reduced, a reward may be given according to the prescribed regulations, and it would be best if over half of the portion of the gasoline so saved is awarded to the unit concerned." He also recommended that maintenance receive special attention: "If the maintenance cost is reduced, then one-third of the saving may be awarded to the unit concerned."[81] This bonus system was reported in operation in Heilungkiang, Kiangsu, Liaoning (which also had penalties), and Shansi.[82] No such bonus system existed in the 1950's.[83]

The system of bonuses in the AMS's seems similar to that of the MTS system in the Soviet Union. A regulation of 1939 specified that premiums would be paid to tractor drivers for exceeding shift norms, fulfilling seasonal norms, and economical use of fuel and lubricants. In addition, the amount of food rations for MTS personnel were related to the actual yield of agricultural production.[84]

Despite these changes some Chinese administrators felt the AMS system was deficient in service and too costly. Hsiang Nan stated bluntly about the situation: "Such a management system has to be changed."[85]

79. "T'an Chen-lin's Crime of Sabotage," SCMM 624, p. 3.

80. "History of Struggle," SCMM 633, p. 32.

81. "T'an Chen-lin's Crime of Sabotage," SCMM 624, p. 3.

82. "Thoroughly Eliminate the Pernicious Influence of Material Incentive," SCMM 605, p. 30; "Refute Material Incentive and Render It Repugnant," SCMM 605, p. 26; Chang Ch'ing-t'ai, "On the Question of Strengthening the Operation and Management of Tractor Stations," *Chung-kuo Nung-pao,* no. 11 (Nov. 10), 1963 (SCMM 400, p. 24); "Tractor Station in Tinghsiang Hsien, Shansi, Deeply Welcomed by Peasants," *Jen-min Jih-pao,* Nov. 26, 1963 (SCMP 3115, p. 12); "Run Tractor Stations Truly Well," *Jen-min Jih-pao* editorial, Nov. 26, 1963 (SCMP 3115, pp. 9–11).

83. Interview with technician who worked in agricultural mechanization.

84. Naum Jasny, *The Socialized Agriculture of the USSR* (Stanford: Stanford University Press, 1949), p. 285. In the early 1950's, bonuses could raise the salaries of station directors, chief engineers and chief agronomists by 50 to 65 per cent (Laird, Sharp, and Sturtevant, *Rise and Fall,* pp. 55, 67).

85. Hsiang, "Agricultural Mechanization Can Be Achieved," SCMP 3518, p. 7.

Advocating and Creating the "Trust"

The new organization proposed in the early 1960's was called "trust" or "corporation." The concept of "trust"[86] comes primarily from Soviet, Western capitalist, and Yugoslavian sources. Liu Shao-ch'i advocated "learning from the capitalist way of managing enterprises, particularly the experience of monopolistic enterprises, and learning from fine Soviet experience."[87]

The Soviet experience was probably primary. Trusts were created in the Soviet Union in October 1921 as the legal entity to manage "commercially autonomous units," i.e., those enterprises which did not require state investment and which did not deliver the bulk of their produce to the state. As a practical matter, these "commercially autonomous units" made up the bulk of the economy, leaving out only military supply industries, municipal enterprises, basic metals, heavy transport, state construction, and similar industries. By the end of 1923, 360 industrial trusts had been established, employing over 80 per cent of the workers employed in state industry. During the 1930's the number of trusts was reduced, as state investment in more industries developed; but "where the unit of production was small . . . or where the industry was complex in structure, factories remained grouped into Trusts, as the responsible financial units, and it was the Trust that had direct connection with the Chief Administration." The trust computed its costs over all its factories; this means that the trust had the discretionary power to reallocate profits among its different factories. The individual plants, however, were not allowed to plan the use of profits resulting from efficient operations.[88]

Another source of inspiration for the trust system was the experience of Western capitalist corporations. This will be apparent from references which appear below.

Yet another set of experiences on which the Chinese drew was the Yugoslavian experiments in "workers' control" over factories. One Red Guard article characterized Yugoslavian policies and charged that Liu Shao-ch'i was interested in them:

86. The Chinese phrase for "trust" is a simple transliteration *"t'o-la-ssu."* The Russian word is *trest,* taken from English.

87. Ching Hung, "The Plot of the Top Ambitionist To Operate 'Trusts' on a Large Scale Must Be Thoroughly Exposed," *Kuang-ming Jih-pao,* May 9, 1967 (SCMP 3948, p. 6).

88. Background on Soviet trusts is drawn from Maurice Dobb, *Soviet Economic Development since 1917* (New York: International Publishers, 1966), pp. 132, 160, 390; the quotation is from p. 371.

> The reactionary program of "autonomy for workers," put forward by the Tito revisionist clique of Yugoslavia . . . resulted in the disintegration and transformation of Yugoslavia's socialist economy based on the system of ownership of the whole people.
>
> In China at present, the top Party person in authority taking the capitalist road [Liu Shao-ch'i] has looked upon Tito as teacher with the object of peddling his black merchandise of capitalist trusts, in a vain attempt to make China take the path of capitalist restoration as Yugoslavia has done.[89]

According to Red Guard charges, Liu Shao-ch'i first proposed the organization of trusts in China in 1960.[90] At an unidentified meeting in October and December 1963 and possibly on other occasions, Liu is reported to have strongly urged the establishment of trusts. Not surprisingly, only fragments of speeches by Liu from 1963 are available.[91] The statements that were released have here been put in an order that attempts to present a coherent argument.

> Our present method is for provinces, municipalities, departments, bureaus, and various departments of the Central Committee to interfere in the economy. This is an extra-economic method; it is not a capitalist method but a feudal method.[92]

> It is necessary to consider the trust method. Control must be exercised over manufacture as well as business management. Rather than set up truck and tractor departments, it is better to organize truck and tractor companies. The operating expenses for agricultural machines should also be controlled by those companies.[93]

> All trades and occupations should operate trusts.[94]

> The bureaus of the various ministries of the central government are to be converted into companies and become not administrative organs but enterprise organizations.[95]

89. Ching, "The Plot," SCMP 3948, p. 3.

90. "T'an Chen-lin's Crime of Sabotage," SCMM 624, p. 6.

91. None of these topics shows up in *Collected Works of Liu Shao-ch'i, 1958–1967* (Hong Kong: Union Research Institute, 1968).

92. Peking Institute of Agricultural Mechanization, Capital Red Guard Congress, East Is Red Commune, "The Reactionary Nature of China's Khrushchev in Promoting the Trust," *Nung-yeh Chi-hsieh Chi-shu,* no. 5 (Aug. 8), 1967 (SCMM 613, p. 20).

93. *Ibid.*

94. Ching, "The Plot," SCMP 3948, p. 3.

95. "Reactionary Nature of China's Khrushchev," SCMM 613, p. 22; also quoted in: Peking Agricultural Machinery College, East Is Red Commune, Criticism and Repudiation Office, "Completely Settle the Heinous Crimes of China's Khrushchev and Company in Undermining Agricultural Mechanization," *Nung-yeh Chi-hsieh Chi-shu,* no. 5 (Aug. 8), 1967 (SCMM 610, p. 26).

Turn departments and ministries into corporations, ministers into managers and vice ministers [into] assistant managers.[96]

The power of the trust is very great, greater than that of the government. As a matter of fact, it runs the government.[97]

The Economic Commission [presumably the central economic planners] should exercise direct jurisdiction over the trust.[98]

[The trust should] take over the supervisory work of the Party exercised by Party organizations.[99]

Trusts are free to purchase raw materials and sell their products. For instance, cotton may be purchased by spinning and weaving concerns, tobacco leaves by cigarette companies, because commercial departments are unfamiliar with this approach which is also the commercial viewpoint but lacks the production viewpoint. Trusts in other countries even have to take care of marketing problems! The Ministry of Petroleum Industry may in the future do something about its marketing outlets.[100]

Trusts may consider doing whatever is profitable.[101]

A factory must make money. Otherwise it should be closed down and wages forfeited. [A factory should go in for] sideline production because this can make more money and more profits.[102]

Some material incentives are needed. If you don't give more money, they [workers and/or managers] will not be enthusiastic in production and work well for you.[103]

In setting up trusts, let us have national ones.[104]

We must create all over the country trusts more monopolistic than those under capitalism.[105]

The socialist economy should be more centralized and more monopolistic than the capitalist economy.[106]

Socialist monopoly will be even greater in degree than capitalist monopolies because it has no competition. . . . The trusts should be on a national, unified basis, free from local interference.[107]

96. Ching, "The Plot," SCMP 3948, p. 3.
97. "Completely Settle the Heinous Crimes," SCMM 610, p. 26.
98. "Reactionary Nature of China's Khrushchev," SCMM 613, p. 22.
99. *Ibid.* 100. Ching, "The Plot," SCMP 3948, p. 5. 101. *Ibid.*
102. "China's Khrushchev Condemned for Pushing Revisionist Line in Economy," NCNA Peking, March 25, 1968 (SCMP 4148, pp. 23–25).
103. *Ibid.* 104. "History of Struggle," SCMM 633, p. 24.
105. "T'an Chen-lin's Crime of Sabotage," SCMM 624, p. 6.
106. Ching, "The Plot," SCMP 3948, p. 3.
107. "History of Struggle," SCMM 633, p. 24.

The question of whether convenience would be accorded to the localities will not be considered.[108]

In short, things must be organized and planned. Don't promote things on your own with no regard for the Center. All local undertakings must be organized, and this is what is called socialism.[109]

By getting organized, it is possible to promote specialization, standardization, and systematization, improve the quality, raise labor productivity, increase variety and lower costs. This will benefit society as a whole, the whole country and all places.[110]

Enterprises are managed by capitalists just the same. Factories managed by capitalists are as a rule well managed. Monopolistic companies are well organized.[111]

The bourgeoisie has hundreds of years of experience in the management of enterprises. . . . Monopoly companies are very well organized internally; they are able to lower costs and raise quality and labor productivity.[112]

Successful bourgeois experience is available [to learn from].[113]

We must study the experience of capitalist management enterprise.[114]

We must manage the economy by economic methods.[115]

Central and local governments [should be] above commercial establishments and above contradictions, [and should be made] arbiters when problems crop up. Local governments should be confined to supervisory functions such as the collection of surtaxes and supervision of municipal projects.[116]

Both Party and Government organizations at central and local levels should be somewhat detached and avoid getting personally involved.[117]

Some additional information about the internal organization of the trusts is known. The provisional regulations for management of the Tractor and Internal Combustion Engine Spare Parts Company and the Tractor and Internal Combustion Engine Industrial Company were said by Red Guard groups to have included these provisions:[118]

108. *Ibid.* 109. "Reactionary Nature of China's Khrushchev," SCMM 613, p. 22.
110. *Ibid.* 111. "Reactionary Nature of China's Khrushchev," SCMM 613, p. 20.
112. Ching "The Plot," SCMP 3948, p. 6. 113. *Ibid.*
114. "T'an Chen-lin's Crime of Sabotage," SCMM 624, p. 6. On the basis of this instruction Po I-po proposed to send a research group to France and other countries to study capitalist practices of industrial management; but the trip fell through (*Hung-se Kung-chiao*, no. 6 [May 12], 1967, p. 1; cited in Peter Nan-shong Lee, "Authority in Chinese Industrial Bureaucracy: Recent Developments," paper for American Society for Public Administration, 1974, p. 17).
115. "T'an Chen-lin's Crime of Sabotage," SCMM 624, p. 6.
116. Ching, "The Plot," SCMP 3948, p. 4. 117. *Ibid.*, p. 2.
118. "History of Struggle," SCMM 633, p. 27.

1. "Retention of profits" at separate levels of the company. Thus an individual factory could use its profits for individual rewards, increased investment, or other purposes. This seems a bit different from the early Soviet model.

2. "Retention of deductible sums of money" at separate levels. The meaning of this is not clear, but perhaps it means that each factory was allowed a certain sum for depreciation; if maintenance were done well, so that the full allocation was not needed for depreciation, then it might be retained and applied to other purposes.

3. "The company takes the responsibility for drawing up a reward system and fixes the rates of reward."

Red Guard groups said that the trust organization was established on the idea that "the highest criterion demanded of an enterprise is 'profit' and the basic principle underlying labor management is 'reward.' "[119] Unfortunately the full text of the provisional regulations for management of the tractor trust is not available, so the Red Guard interpretation cannot be evaluated.

While the trust was envisioned as a centralized bureaucratic organization controlled by economic criteria, it differed in several important ways from a corporation in a capitalist country. There is no evidence that the trust was free to establish prices or to raise investment capital independently of government economic plans, although it may have been free to invest its profits. Thus its production plans could have had only limited flexibility. Finally, a single trust in each sector would form a monopoly, so that no competition would emerge. A commune would not be able to shop and compare the services of different tractor stations.

Twelve trusts were established and others contemplated. The first, to be discussed in detail below, for (1) agricultural machinery, was set up in 1963. During 1964 and 1965 trusts were established in the (2) pharmaceutical, (3) salt, and (4) tobacco industries. During 1966, pilot programs for establishing trusts were developed in the (5) petroleum, (6) textile, (7) automobile, (8) steel, (9) coal mining, (10) hydroelectricity, and (11) river navigation industries.[120] Trusts were also set up (or planned) in the (12) rubber and (13) aluminum industries.[121]

119. *Ibid.*

120. "T'an Chen-lin's Crime of Sabotage," SCMM 624, p. 6. These trusts are specified by Lee, "Authority in Chinese Industrial Bureaucracy," p. 18.

121. Lowell Dittmer, *Liu Shao-ch'i and the Chinese Cultural Revolution* (Berkeley: University of California Press, 1974), pp. 249–250.

The origins of the agricultural machinery trust go back to 1963, when the Peking Municipal Agricultural Machine Corporation integrated the production plans of various factories in Peking related to agricultural machinery.[122] In addition, it was reported that the Number 1 Tractor Factory located at Loyang had launched a campaign to increase production and profits (as well as to improve quality and labor productivity and reduce expenses). This suggests that profit was an important criteria for the performance of a factory.[123]

At a December 1963 meeting, Liu Shao-ch'i gave instructions that a trust should be formed in the agricultural mechanization departments:

> It is good to have agricultural machine supply centers. A big trust should be formed and supply substations should be set up along railroads and highways. Don't set up stations according to administrative districts and don't put them under the direct jurisdiction of counties. Local authorities must not lay their hands on such stations. They can make suggestions, but cannot allocate money for making such stations. All agricultural machines should be under the unified management of the supply company and factories should also be under its control. Tractors, irrigation supply company and factories should also be under its control. Tractors, irrigation and drainage machines and oil supply should be under the unified management of the company.[124]

Very likely at the same meeting, T'an Chen-lin stated experiments with trusts would be undertaken.

> According to the directive of Comrade Liu Shao-ch'i that the supply of (agricultural) means of production should be handled by farm machinery companies . . . experimental points will be established next year and the work is expected to be completed in two or three years. Effort will be directed first at machinery and oil-bearing crops and then at farm insecticides and chemical fertilizer.[125]

One of the experimental trusts was established on a regional basis in Shensi Province, in one administrative district (Weinan), and confined there for about a year.[126] The experiment was claimed to be successful: "Since the last one and a half years, the company has harmonized indus-

122. "News About Agricultural Machines," *Chung-kuo Nung-yeh Chi-hsieh*, no. 6 (June 10), 1963 (SCMP 374, p. 9).

123. *Ibid.*

124. "Reactionary Nature of China's Khrushchev," SCMM 613, pp. 20–21.

125. "T'an Chen-lin's Crime of Sabotage," SCMM 624, p. 6.

126. Shensi Bureau of Agricultural Machinery, Revolutionary Leading Group "T'an Chen-lin Cannot Shun His Criminal Responsibility for Closely Following China's Khrushchev in Promoting Capitalist Trust," *Nung-yeh Chi-hsieh Chi-shu*, no. 9, 1968 (SCMM 644, p. 35).

trial-commercial relations, balanced production-supply plans, simplified administration, promoted the adoption of semi-mechanized tools and new products, and greatly strengthened agricultural production.''[127] By early 1965, it had been expanded to include three administrative districts, and by the end of 1965 it covered the entire province. It eventually controlled 120 local factories and stations, 6 major factories, 4 schools, 10 research stations, and about 100 stores.[128]

One of the particularly interesting features of the experimental trust in Shensi was that it ėstablished a new style of Communist Party organization. In factories that were part of the trust, the Party branches were primarily subject to the leadership of Party committees in higher-level factories. The instructions for setting up the system in Shensi advocated that a ''Party committee be set up at the provincial company, and the Party committees attached to district, county, or municipal companies should be led by the company Party committees of a higher level and the local Party committees, but principally by a higher Party committee of the company.''[129] Red Guards accurately perceived that this pattern of party organization was similar to the ''industrial party'' advocated by Khrushchev. P'eng Chen made the same observation, although it is not clear whether in criticism or support: ''When we operate a trust and have the trust take over the work of the Party, we are in fact running an industrial party.''[130]

Another experimental trust was established in March 1964, when the China Tractor and Internal Combustion Engine Spare Party Company was set up with control over 13 local plants making tractor spare parts.[131] The authority of the trust grew, and by 1965 it controlled the distribution of 5,000 kinds of spare parts.[132] By early 1965, the spare-parts trust was judged successful, and expansion was urged. Po I-po is quoted as saying,

127. ''Report on the Experimental Wei-nan Semi-Mechanized Agricultural Tool Company,'' *Chung-kuo Nung-yeh Chi-hsieh,* no. 7 (July), 1965, pp. 19–23 (JPRS 42,484, p. 47).

128. ''T'an Chen-lin Cannot Shun His Criminal Responsibility,'' SCMM 644, p. 35.

129. *Ibid.,* p. 37.

130. In the course of Mao's talk at Enlarged Meeting of the Political Bureau, March 20, 1966, *Miscellany,* p. 379.

131. ''History of Struggle,'' SCMM 633, pp. 24, 27; ''T'an Chen-lin's Crime of Sabotage,'' SCMM 624, p. 7; ''News About Agricultural Machinery,'' *Chung-kuo Nung-yeh Chi-hsieh,* no. 4 (April), 1965, pp. 2–3 (JPRS 42,484, p. 10); ''Completely Settle the Heinous Crimes,'' SCMM 610, p. 27.

132. ''Completely Settle the Heinous Crimes,'' SCMM 610, p. 27.

in an unidentified statement in May 1965, "It won't do to set up a spare parts company alone."[133] In July, the leadership of the Eighth Ministry of Machine Building (which supplied agricultural machinery and oversaw the new spare-parts trust) reported "wholehearted support for and determination to act on the instructions" of two top Party "capitalist roaders" to broaden the spare-parts trust. They agreed that "a Tractor and Internal Combustion Engine Company must be set up promptly and run properly."[134]

Thus, on August 1, 1965, the China Tractor and Internal Combustion Engine Industrial Company was officially inaugurated.[135] The new company was expected to absorb more than a hundred local enterprises and had eight regional branches, to operate in the Northeast, Shanghai, Tientsin, and other places.[136]

The Debate over Expanding the Agricultural Machinery Trust

As soon as the trust was expanded from spare parts to manufacture of engines, another even more dramatic expansion was proposed. Plans were drawn up—including an "Organizational Outline for Agricultural Machinery Companies in China"—for the creation of the China Agricultural Machine Company. This company would not only manufacture agricultural machinery and spare parts but would also control repair shops, research institutes, and—most importantly—the network of agricultural machinery stations. Branch companies in the provinces, municipalities, and autonomous regions, and companies at the county level were contemplated.[137]

An integral element of the plan was that the proposed trust would concentrate its efforts in roughly 100 counties (out of China's more than 2,000). A conference was held in October 1965 by the Eighth Ministry of Machine Building to draw up a plan for the selection of 100 key counties.[138] Actually, 130 counties were selected. Various leaders explained

133. "History of Struggle," SCMM 633, p. 24. 134. *Ibid.*

135. "Agricultural Machinery News," *Chung-kuo Nung-yeh Chi-hsieh,* no. 9 (Sept.), 1965, pp. 2–3 (JPRS 42,545, p. 77).

136. "History of Struggle," SCMM 633, p. 24; "Reactionary Nature of China's Khrushchev," SCMM 613, p. 21.

137. "Reactionary Nature of China's Khrushchev," SCMM 613, p. 21; "Completely Settle the Heinous Crimes," SCMM 610, p. 27; "T'an Chen-lin's Crime of Sabotage," SCMM 624, p. 7.

138. Within the state farm system, a "keypoint mechanized production team" program had been adopted in 1963. In these keypoint teams, machinery was concentrated to win "battles of annihilation," and to serve as "vanguards in mechanization." One hundred

the logic of concentration of resources. Liu Shao-ch'i said, "We must equip whole counties one by one and must not equip them on a piecemeal basis."[139] An unidentified leader stated: "In the past, tractors were scattered and there was no war of annihilation. . . . Let us not scatter pepper. . . . Counties must be equipped one by one."[140] P'eng Chen agreed: "Use of machinery must be centralized. . . . If ten or eight tractors are allocated to one county they cannot be well maintained. Tractors must be used in a centralized manner in counties one by one."[141]

The 130 counties slated to receive mechanization were not listed publicly, but a Red Guard group in Heilungkiang indicated the criteria used in the selection:

> In regard to the distribution of tractors, the capitalist roaders laid emphasis on counties with plenty of land but short of draft animals, with land to reclaim and a high percentage of commodity grain. They regarded agricultural mechanization as a provisional measure for "stopping the crack," "filling the gap," "digging into the purse," "reaping ready profit," and improving desolate and barren land.[142]

These criteria would, of course, mean that the newly proposed trust would operate primarily in Northeast and North China.

The proposed Agricultural Machine Company would be financially self-sustaining after initial capitalization. A directive sent (by Liu Shao-ch'i, according to Red Guard groups) to Heilungkiang detailed the financial plans:

> In ten years, the State will invest a sum of money and it will begin to recover it after ten years. The state will use this sum of money to equip another area and complete mechanization is to be realized in several decades. The state will have

---

such keypoints were set up in Heilungkiang, with 1.5 million mou. The yield on this land was 40 percent higher than in unmechanized areas, and its commodity rate was 80.5 per cent. The plan to develop 100 key counties was undoubtedly related to this previous experiment in the state farm system. Shang Chih-lung and Ma Ching-p'o, "Fifteen Years of Agricultural Mechanization on State Farms," *Nung-yeh Chi-hsieh Chi-shu,* no. 11 (Nov. 13), 1964 (SCMM 451, p. 7).

139. "Completely Settle the Heinous Crimes," SCMM 610, p. 30; also quoted in Heilungkiang Department of Agricultural Mechanization, Mao Tse-tung Thought Red Guards and the Red Rebel Regiment, "Thoroughly Wipe out the Pernicious Influence of the Bourgeois Reactionary Line, Let the Great Red Banner of Mao Tse-tung's Thought Be Planted all over the Agricultural Mechanization Front," *Nung-yeh Chi-hsieh Chi-shu,* no. 2–3 (May 23), 1967 (SCMM 588, p. 11).

140. "Completely Settle the Heinous Crimes," SCMM 610, p. 30.

141. *Ibid.*

142. Heilungkiang Department of Agricultural Mechanization, Committee Taking over the Control, "Let the Radiance of Mao Tse-tung's Thought for ever Shine over the Road of Agricultural Mechanization," *Nung-yeh Chi-hsieh Chi-shu,* no. 4 (July 8), 1967 (SCMM 600, p. 2).

to invest ten to twenty billion *yuan*. In the future, we shall have to rely on these sums of money to do our work.[143]

It would appear that the state planned to invest one or two billion yuan each year for ten years in the agricultural machinery trust. After that, the branches of the trust would have to make a profit, and the profit would be reinvested in other counties.[144] The trust would, presumably, try to place stations only in those locations where they would show a profit, so that additional funds would be generated for expansion. In such a financing scheme, the agricultural production units, including communes, brigades, or teams, would not have to raise cash to purchase agricultural machinery because they would rent the services. Although Red Guards made no analysis of the adequacy of the proposed investment fund of ten to twenty billion yuan, it was quite substantial, if not fully adequate.[145]

During the summer of 1965 efforts were made to get this expansion plan for the trusts approved by the Central Committee. In July there was a National Conference on Agricultural Machine Work for directors of Agricultural Machinery Management Bureaus at which P'eng Chen made a major report.[146] In August, the Eighth Ministry of Machine Building convened a National Agricultural Machinery Management Conference. From the little information that is available, it seems that this meeting was sharply divided on the issue of the expansion of the trusts. Fragmentary quotations from Red Guards indicate some support for the trusts. An unidentified spokesman said: "Agricultural machinery supply stations are good. Let us set up a trust . . . set up supply branch stations along

143. "Completely Settle the Heinous Crimes," SCMM 610, p. 30; also quoted in "Let the Radiance of Mao Tse-tung's Thought," SCMM 600, p. 2.

144. "Two Diametrically Opposite Lines," SCMM 633, p. 42.

145. Ten billion yuan could supply one million standard tractors costing ¥10,000, roughly what had been estimated to fill the national need. This would leave another ten billion yuan (if allocated) for implements, repair shops, construction, and training personnel. This was enough to make a major dent in the national mechanization program. Another way of looking at the proposed budget is to think of it as representing roughly ¥130,000 – ¥260,000 per commune. Again, this is substantial but not enough to do the full job of mechanization. One commune in Hupeh spent ¥1,500,000 for mechanization (Hsiang, "Agricultural Mechanization Can Be Achieved," SCMP 3518, p. 4). Another commune spent ¥840,000 ("Commune on Yangtze River Island Archives Mechanized Farming by Its Own Efforts," NCNA Nanking, July 12, 1966; SCMP 3740, p. 21). A commune in Hupeh spent ¥1,200,000 ("China's Farm Machinery Industry Grows by Leaps and Bounds," *Peking Review,* no. 40 [Sept. 30], 1970, p. 20).

146. "T'an Chen-lin's Crime of Sabotage," SCMM 624, p. 6; Eighth Ministry of Machine Building, United Committee of the Revolutionary Rebels, "57" United Detachment, "Wipe Out State Monopoly and Promote Mechanization on the Basis of Self-Reliance in a Big Way," *Nung-yeh Chi-hsieh Chi-shu,* no. 6 (Sept. 18), 1967 (SCMM 610, p. 14).

railway lines and highways. We should not set up stations according to administrative divisions nor should we make direct allocation to the county. And all stations should be run independent of the local government."[147] But the conference as a whole opposed the trust system and looked toward commune ownership as the eventual solution to the contradiction between the state-owned machines and the cooperatively owned land. It was concluded: "In localities where conditions were favorable, communes should be encouraged to use collective accumulation to buy farm machinery and rely on collective strength in operating farm machinery."[148] The conference adopted a guideline of "integrating stations with communes and State-operated stations with collective-operated stations but regarding collective operation as the leading factor."[149]

The campaign to expand the trust system was not ended by this conference, however. A meeting in October selected the keypoints for the expanded China Agricultural Machine Company. Moreover, Liu Shao-ch'i traveled around Hopei and to Heilungkiang and other provinces to explain and advocate the trust system.[150] He thought the understanding and support of people affected by a policy would be important in having the policy accepted.

What remains unclear is whether the political system made an authoritative decision to proceed on the expansion of trusts. Liu believed that the proposal to establish the enlarged Agricultural Machine Company had been officially accepted, and apparently sent directives to Heilungkiang urging implementation of the plan. He seems to have considered his trips to have been part of setting up the new system. Moreover, there is weak

147. "Completely Settle the Heinous Crimes," SCMM 610, p. 27.

148. "Opening up the Road of Agricultural Mechanization in Our Country, Proceeding from the Needs of 500 Million Peasants and the Characteristics of People's Communes," *Jen-min Jih-pao,* Aug. 31, 1965 (SCMP 3543, p. 8). This article is a summary of the August meeting, published before the Cultural Revolution. A brief summary was released also: "National Conference on China's Agricultural Mechanization," NCNA Peking, Aug. 29, 1965 (SCMP 3531, p. 11). The article about the August meeting was accompanied by a *Jen-min Jih-pao* editorial: "Management of Agricultural Machinery Should Better Serve Agricultural Production," Aug. 31, 1965 (SCMP 3543, p. 9).

149. "Opening Up the Road of Agricultural Mechanization," SCMP 3543, p. 8.

150. "Wipe Out State Monopoly," SCMM 610, p. 14. His explanations of the system during his tour of Heilungkiang were dutifully recorded by agricultural cadres, and became the basis for attacks on Liu during the Cultural Revolution. These records also provide us with many of the above quotations. The best example is the article by the Committee Taking over the Control of the Department of Agricultural Mechanization of Heilungkiang, "Let the Radiance of Mao Tse-tung's Thought," SCMM 600, pp. 1–7; another which incorporates some of the material that Liu said in, or wrote to, Heilungkiang is "Completely Settle the Heinous Crimes," SCMM 610, p. 30.

evidence that the idea of trusts was being incorporated into the Third Five-Year Plan. A *People's Daily* editorial published on January 1, 1966, entitled "Welcoming 1966—the First Year of China's Third Five-Year Plan," included slogans which had been used to describe trusts: "We shall have to speed the construction of new projects by way of 'waging a war of annihilation with concentrated forces' in order to enable them to go into early operation. We shall organize the circulation of commodities rationally and raise the standard of the trading services so as to better serve production and the people."[151]

Red Guard groups insisted, however, that the Central Committee never accepted the plan. They say that the sponsors of the plan to set up the Agricultural Machine Company "repeatedly urged Premier Chou to approve it,"[152] but "due to the strong opposition of the revolutionary masses," this plan could not be effected.[153] Perhaps Liu felt that such a policy could be implemented by the concerned departments without formal approval of the Central Committee, where Mao could block it.

The question of trusts appears to have been debated openly for the last time at the National Industrial and Communication Work Conference sponsored by the State Economic Commission and the Political Department of Industry and Communications of the Central Committee in late February and early March 1966. Contemporaneous documents reveal widespread opposition to Mao's policies at the conference: "A complete system of enterprise management has still to be worked out and only a part of our cadres understand how to apply Mao Tse-tung's thinking in running enterprises."[154] An unidentified "capitalist roader" supported the idea of trust, saying, "In promoting the trust . . . the [general] orientation is correct."[155]

Red Guard sources report Mao's emphasis on self-reliance in his recommendation to "various provinces, municipalities and regions to map out a five-year plan, seven-year or ten-year plan on the basis of self-reliance, to expand gradually from a few experimental spots and to spend

151. Translated in *Peking Review,* no. 1 (Jan. 1), 1966, p. 7.
152. "T'an Chen-lin's Crime of Sabotage," SCMM 624, p. 7.
153. "The Reactionary Nature of China's Khrushchev," SCMM 613, p. 21.
154. *Ibid.*
155. China Institute of Research in Agricultural Mechanization, Capital Workers' Congress, East Is Red Commune, "Last-Ditch Struggle that Courts Self-Destruction—Denouncing the Towering Crimes of China's Khrushchev in Opposing Chairman Mao's Wise Decision," *Nung-yeh Chi-hsieh Chi-shu,* no. 5 (Aug. 8), 1967 (SCMM 613, p. 25).

twenty-five years to realize in the main the mechanization of agriculture."[156]

On March 12, when the conference was still in session,[157] Mao sent out a letter replying to a letter from Liu on proposals to mechanize Hupeh. (See below, p. 224.) The letter included a critique of the trusts. Fragments of it say:

> [Agricultural mechanization] must be carried out in the main by various provinces, municipalities and regions on the basis of self-reliance, and the Center can only give assistance in the form of raw and semi-processed materials to places short of such materials.
>
> Local authorities must be given the right to manufacture some machines. . . . It is not a good way to exercise too rigid a control by placing everything under the unified control of the Center.
>
> [Agricultural mechanization] should be linked with making preparations for war, for famine and for the people, otherwise it will not be carried out with enthusiasm in places with the necessary conditions.
>
> Agricultural mechanization must be linked with these aspects before the masses can be aroused to fight for the faster mechanization by steadfast realization of this kind of planning.
>
> The agricultural policy of the Soviet Union has always been wrong in that it drains the pond to catch the fish and is divorced from the masses, thus resulting in the present dilemma. . . . We should at least take warning from it. . . . Has not the mechanization of agriculture been realized in the main in the Soviet Union? What has led to its present dilemma? This is food for thought.[158]

Another indication of Mao's displeasure was revealed a few days after he sent this letter, at a March 20, 1966, meeting of the Politbureau. When informed by Chou En-lai that "the Eighth Ministry of Machine Building operated a trust and took over quite a number of plants," Mao pungently replied: "Then tell XXX of the Eighth Ministry of Machine

156. "Wipe Out State Monopoly," SCMM 610, p. 16.

157. The conference continued at least until March 11, because on that date Lin Piao sent a letter to the conference (published in *Jen-min Jih-pao* on June 19, 1966). Philip Bridgham, "The Great Proletarian Cultural Revolution," *China Quarterly,* no. 29 (Jan.–March 1967), p. 20 n. 33.

158. These fragments are all available in "Last-Ditch Struggle," SCMM 613, p. 26. Some are also available in "Wipe Out State Monopoly," SCMM 610, p. 16; and "The Reactionary Nature of China's Khruschchev," SCMM 613, p. 21. The full text of Mao's Letter of Instructions on the Question of Agricultural Mechanization, March 12, 1966, is available in *Miscellany.,* p. 373, and *Peking Review,* no. 52 (Dec. 26), 1977, pp. 7–9.

Building to go and become the plant director!''[159] A summary of part of the conference was published as an editorial on April 3, and suggested that the Maoist position on many questions—such as the importance of the "mass line"—had been confirmed.[160]

This period of the early 1960's raises a profound question which continues to shape China's political debate in the 1970's and will undoubtedly remain important for decades: What patterns of social organization are needed to achieve technological reform? Some of China's top leadership in the early 1960's, including Liu Shao-ch'i, Teng Hsiao-p'ing, and P'eng Chen, believed that technological reform required efficiency; that efficiency in turn required centralized administration, specialization of functions in factories and bureaucracies, and material incentives for workers and administrators; and that profits were the crucial measure of efficiency. In short, they believed that the Soviet model, perhaps modified marginally, was required to modernize China's economy. They may have admired and respected Mao Tse-tung and his values, but they believed that his visions—notions of social integration; of reducing differences between agriculture and industry, countryside and city, manual and mental labor; of spiritual development and nonmaterial incentives—were not compatible with a modern technological society. They started to implement centralized, specialized patterns of political and social relationships in the early 1960's. Mao and others sharply disagreed. Mao created the Cultural Revolution to force discussion of these issues and to convince the Chinese people that modernization was compatible with revolutionary values—and if, perchance, it was not, then the revolutionary values were more important. The next chapter will examine the programs suggested during the Cultural Revolution; but it should be noted here that the Cultural Revolution did not extinguish the notion that modernization required centralization, material incentives, specialization, and profits. Again in 1975 it appeared that Teng Hsiao-p'ing put forward a program for economic development that had many similarities to the proposals of the early 1960's.[161] He was purged and replaced by Hua Kuo-feng, only

159. Mao's talk at Enlarged Meeting of the Political Bureau, March, 20, 1966, *Miscellany,* p. 380.

160. "Running Enterprises in Line with Mao Tse-tung's Thinking," *Jen-min Jih-pao* editorial, April 3, 1966, available in *Peking Review,* no. 16 (April 15), 1966, p. 11.

161. In 1975, Teng Hsiao-p'ing drafted several documents outlining a new economic strategy. Three of them are entitled "On the General Program for All Work of the Party and the Country," "Some Problems Concerning the Work of Science and Technology" (called "An Outline Report" for short), and "Some Problems in Accelerating Industrial Develop-

to re-emerge again in the fall of 1977. It is inconceivable that Teng was the only person who held these ideas and that he put together the program single-handedly in the early 1960's. The notion of centralized, specialized organization of the early 1960's undoubtedly continues to have widespread support in China and most likely will be recommended from time to time in the future.

The development of agricultural mechanization policies in the 1960's reveals important aspects of the decision-making process in China. While a few individuals (such as Liu Shao-ch'i, P'eng Chen) played a crucial role in proposing new programs, they did not work in a vacuum. They had to mobilize fairly broad support among the concerned departments by meeting and traveling and explaining their proposals at the local level. Most remarkably, there was some ambiguity within the system concerning the making of decisions. Liu Shao-ch'i and others thought that decisions could be made by ministries and departments. Perhaps they could have been—if Mao Tse-tung had not objected. But Mao forced the issue to different political arenas in the course of the Cultural Revolution. The importance of Mao to the decision-making process is crucial, as will be apparent in the following chapter. His death creates a hole in the Chinese polity; significant changes must develop in the political process.

---

ment." While the texts have not been released, criticisms of them indicate that they contained programs similar to those of the early 1960's, particularly the trust. Criticisms of these programs are available in several issues of *Peking Review* during 1976.

9. The Cultural Revolution: Decade of Mechanization

During the Cultural Revolution the priority of agricultural mechanization was reaffirmed, precisely because agricultural mechanization affects the whole social structure so profoundly. Moreover, mechanization was placed in the hands of the communes; the power of the state bureaucracy, in the form of AMS's or trusts, was diminished.

When the Cultural Revolution broke out in 1966, it took the world by surprise. In retrospect, the Cultural Revolution seems inevitable. Although the first manifestation of the Cultural Revolution involved criticisms of theater and education, the revolution was by no means confined to cultural questions. Indeed, cultural questions were a starting point for political debate between alternative visions of a socialist future, in which the role of agricultural mechanization and the position of the rural sector would play a significant part. The visions had been debated for years. One, based on the Soviet Union's experience, was partially implemented in China during the First Five-year plan. This vision of socialism stressed the importance of building a powerful heavy-industry sector, principally in a few urban and regional centers. Strong, centralized government was desired. The emergence of an elite controlling the economy, basing its power both on governmental power and scientific-technical expertise, was considered appropriate. In this model of development, the rural sector is exploited, and peasants are left in a little-changed situation. They eventually may participate in modernization only by being drawn into the urban-industrial labor market. As long as they remain in the rural sector, they remain in the twilight of modernization.

The other vision of modern socialist society stressed that transformation must take place within the rural sector. Fruits of agricultural modernization should not be extracted from the countryside to supply capital for urban-industrial development. Rather, they should remain in the rural sector, to provide improved nutrition, health, and education services, and to provide investment funds for rural industrialization (and those urban industries which could supply agricultural inputs). In this vision of modernization, the greatest risk is the emergence of an elite in either politics or economics which might capture most of the benefits of modernization and block improvements in the conditions of the great majority. This vision underlay the Great Leap/communization policies of 1958–60, but the technical foundations did not yet exist for this policy to work.

By the mid-1960's, this debate had to come up again, precisely because of economic success in previous years. During the First Five-Year Plan (1953–57) the foundations for modern industry had been established. From 1960 to 1966 this industrial base had been oriented to meet the needs of agriculture. Mechanical irrigation systems were installed; chemical fertilizer was supplied in increasing amounts; high-yield varieties of rice and wheat were developed; various types of farm machinery were designed. By 1963 the agricultural crisis was over, and from 1964 to 1967 grain production increased at the very rapid rate of 5 to 6 per cent per year.[1] The struggle between the two broad visions of socialist modernization could be resumed.

Because of the immense implications of these alternative strategies, the clash took place in virtually every nook and cranny of life in China. It was, however, far more complicated than a debate over two alternative visions. Personal and bureaucratic interests, complicated by misunderstandings and provocations, provided additional dimensions.[2] Here only one aspect of the struggle will be examined, namely the debate over agricultural mechanization.

When the Cultural Revolution was over in 1968–69, China was strongly committed to a program of agrotechnical reform, emphasizing (but not limited to) mechanization. The 1970's can be considered the decade of mechanization of China's agriculture. From an institutional point of view, most machinery is owned and operated by the comunes. It is not used to consolidate the power of the center or of the state bureaucracy. Some of these principles appear to have been challenged by Lin Piao in 1971, but basically in the period since 1967 China has placed great em-

1. For a fuller discussion see Benedict Stavis, *Making Green Revolution* (Ithaca: Cornell Rural Development Committee, 1974), p. 12.

2. The literature on the Cultural Revolution is immense, but two books that effectively uncover the way in which political values intertwined with personal, bureaucratic, and power interests are Jack Chen, *Inside the Cultural Revolution* (New York: Macmillan, 1975), and David Milton and Nancy Dall Milton, *The Wind Will Not Subside* (New York: Pantheon, 1976).

phasis on all phases of agrotechnical reform, and on mechanization in particular.

Issues in the Cultural Revolution

The Priority of Mechanization

There were several reasons why the issue of agricultural mechanization became more salient by the mid-1960's. The country had recovered from the agricultural crisis of 1960–62, so that long-term questions could be considered and major investments could be made. Moreover, from 1960 to 1964 the agricultural machinery industry had advanced significantly. A wide variety of machines and implements, designed specifically for Chinese agriculture, were available, as were spare parts and repair facilities. In addition, rapid increases in oil production, as exemplified by the opening of the Tach'ing complex, increased the availability of fuel for agricultural machines. A strict management system, adopted after 1961, assured full utilization of existing equipment. While the problems were not completely solved, for the first time there were no critical bottlenecks, no major crises in the agricultural machinery industry, and therefore an opportunity for deliberate choice of policies existed.

Perhaps the most important reason for a re-evaluation of the role of machinery in 1964 may have been the social trends during late 1961 and 1962. During the agricultural crisis, the collective economy disintegrated somewhat. Kang Chao suggests that private and family farming became dominant in China in 1961–62.[3] Dick Wilson thought that in 1962 private farming exceeded collective farming in Yunnan, Kweichow, and Szechwan.[4] It is impossible to know the extent of private and family farming in 1961–62 and much data show the collective system did not break down.[5] More likely, cadres in different localities tolerated different systems on a temporary basis, waiting to see how production would be

3. Kang Chao, *Agricultural Production in Communist China, 1949–1965* (Madison: University of Wisconsin Press, 1970), p. 64.

4. Dick Wilson, "The China After Next," *Far Eastern Economic Review* 59:5 (Feb. 1, 1968), 193.

5. This summary is based on several issues of *Pao-An Bulletin* from the fall of 1961, available in *Union Research Service* 27:7 (1962), 111–139. Other information is available in Wang Hung-chih, "Implementation of the Resolutions of the Tenth Plenum of the Eighth Central Committee on Strengthening the Collective Economy and Expanding Agricultural Production," available in C. S. Chen and Charles Ridley, *Rural People's Communes in Lien-Chiang* (Stanford: Hoover Institution Press, 1969), pp. 37, 99; and Jan Myrdal, *Report from a Chinese Village* (New York: Pantheon, 1965).

influenced, and waiting for instructions from the Central Committee to clarify the situation.

Some Central Committee officials (perhaps Liu Shao-ch'i, certainly T'ao Chu) were interested in the family responsibility system, made tests, and suggested it at the Tenth Plenum.[6] The Tenth Plenum rejected the idea of individual or family farming and, at Mao's insistence, stressed collective farming and class struggle.[7] At the same time, however, some compromise was reached; the production team was made the basic accounting unit, small private plots were again guaranteed, the socialist principle of payment according to labor was again emphasized, and family sideline activities were again encouraged.[8] During 1963 the socialist education campaign began, with the objective of strengthening the collective socialist economy.

While the tendency for reintroduction of household farming was beaten back in 1962 and kept back in 1963, it was advocated by some of the top leadership. Basic farm technology in most places had not advanced to the point where collective farming was necessary to utilize inputs on a large scale. Some leaders (such as Mao) must have continued to see in agricultural mechanization the way to assure that that collective economy would be profitable.

As in the past, the question of widespread mechanization of agriculture, especially in the South and Southwest, came back to the fear that labor would be displaced in the densely populated areas and unemployment would result. Hsiang Nan pinpointed the debate in a series of articles published in March 1965: "In areas with large tracts of land and a small population—such as Sinkiang, Inner Mongolia, and Heilungkiang—the controversy over the introduction of mechanization does not arise. However, in areas with little land and a large population, views are by no means unanimous."[9]

6. "10,000 li East Wind" of Red Flag of Sun Yat-sen University, Red Headquarters of Canton Combined Committee for Criticism of T'ao Chu, "T'ao Chu is the Vanguard in Promoting 'Production Quotas Set at Household Level' of China's Khrushchev," *Nan-fang Jih-pao,* July 26, 1967 (SCMP 4011, p. 16).

7. "Communique of the 10th Plenary Session of the 8th Central Committee of the Chinese Communist Party," NCNA Peking Sept. 28, 1962 (CB 691, p. 4).

8. "Revised Draft Articles for the Operation of the People's Communes," Sept. 1962, Articles 39–41 and 20, available in Union Research Institute, *Documents of Chinese Communist Party Central Committee,* Vol. 1 (Hong Kong: Union Research Institute, 1971), pp. 695–725.

9. Hsiang Nan, "Stable and High Yields and Agricultural Mechanization," *Jen-min Jih-pao,* March 22, 1965 (SCMP 3436, p. 12; see also p. 10), and "Agricultural Mechanization

To gather information on agricultural mechanization in general and on the question of displacement of labor in densely populated areas in particular, Hsiang Nan and others conducted a five-month investigation during late 1963 and early 1964. The tour included visits to regions of extensive cultivation in Heilungkiang, Liaoning (Fuhsin County), Inner Mongolia, and Peking. The investigation also examined mechanization of irrigation and processing in Hupeh (Hsinchou County and O-ch'eng County, both in the densely populated lakes region east and northeast of Wuhan), Hunan (Hoyang County), Kwangtung (Nanhai and Tungkuan counties, both fairly near Canton in the Pearl River Delta), and Kwangsi. Roughly half of Hsiang Nan's report was devoted to proving that mechanization was a proper policy in these densely populated areas.

Hsiang based his conclusions on a detailed study of two locations: Hsinchou in Hupeh and Nanhai near Canton. (Both localities had received much machinery to test and demonstrate its utility.) The report of this trip is summarized in Table 9.1. In both counties, mechanization contributed not only to increased labor productivity but also to increased yield.

What happened to people who previously had labored to irrigate the land? In place, the labor force was redirected to three other tasks: intensification of cultivation, diversification, and capital construction.

> After mechanization was introduced, how did Hsinchou and Nanhai handle their labor power thus saved? They promptly organized their labor power for more activities of intensive farming such as plowing, hoeing, tilling, accumulating manure, selecting seeds, and preventing insect pests. Laborers were organized to go up to the mountains or down to the river—for purposes of undertaking afforestation, animal husbandry, fishing, and developing a diversified economy. They were also organized for activities of capital construction on farms such as cutting canals, building roads, building reservoirs and dams, and levelling land. They not only have not faced the problem of "surplus" labor power, they still feel a strain on labor power.[10]

That labor power would be utilized effectively was not, of course, automatic. This was a political question and required correct leadership. Hsiang Nan noted that some places lacked the ability to shift labor to other tasks.

Can Be Achieved with Good and Fast Results," *ibid.*, July 6, 1965 (SCMP 3518, p. 3). The *Jen-min Jih-pao* series was edited from an article by Hsiang in *Chung-kuo Nung-yeh Chi-hsieh*, May 1965, pp. 11–20, entitled "An Inspection Report on the Mechanization of Our Agriculture," JPRS 33,691, pp. 14–29.

10. Hsiang, "Stable and High Yields," SCMP 3436, p. 14.

Table 9.1. Partial summary of Hsiang Nan's report, 1965

Characteristic	Hsinchou County, Hupeh	Nanhai County, Kwangtung
Location	55 km. NE of Wuhan, near lake	15 km. SW of Canton, in Pearl River Delta
Density	1.5 mou/capita (100 people/ha.)	1 mou/capita (67 people/ha.)
Crops	2 cotton, 1 wheat	Paddy rice
Machinery available	Electric irrigation and drainage equipment, 700 water pumps, 500 threshers, 1,400 cotton gins, 700 grain processing machines, 70 oil presses, 64 standard tractor units.	362 irrigation/drainage stations, 148 standard tractor units, 30 diesel engines, 490 water pumps, 28,000-kilowat equipment.
Influence on production	Cotton output up 56%. Now self-sufficient in grain. Bumper harvest in oil-bearing crops, live hogs, fisheries, tobacco, lotus seeds.	Rice output up 48% despite drought.

Source: Hsiang, "An Inspection Report on the Mechanization of Our Agriculture," JPRS 33,691, pp. 14–29.

The phenomenon of "surplus" labor power indeed exists. This is because [some] people have merely used mechanization to reduce labor intensity and raise labor productivity. Thus they may afford to have more "idle" moments because mechanization merely aims at "using a lamp without oil and tilling land without cattle." They do not march on the depth and breadth of production. As a result, mechanization does not bring them the advantage of increased output and a higher income. Instead they have incurred an extra item of expenditure for the purchase of machinery and created "surplus" labor.[11]

Hsiang concluded that agriculture everywhere in China needed mechanization. Indeed, he recalled Mao's statement, "Only when all branches of production and all places where work can be done by machinery are using it, will the social and economic appearance of China be radically changed."[12] Hsiang pointed out that Mao "refers to 'all' branches and places, and not to those places with little manpower but large tracts of land."[13]

While Hsiang wholeheartedly recommended mechanization, he insisted that it be done carefully. He stated, "In achieving mechanization . . . there is no fixed formula. Any decision must be based on actual conditions

11. *Ibid.*, p. 15. 12. *Ibid.*, p. 13. 13. *Ibid.*

in a given place at a given time." For broad guidelines, he suggested the following:

> Generally speaking, in places where population is small and land vast—such as our Northeast, Northwest, the eastern part of Inner Mongolia, etc.—we may start with mechanization of cultivation and of transportation. In those areas where the population is large and land is scarce—such as South China—it may be more realistic if we start with mechanization of drainage and irrigation and of processing farm and subsidiary products.[14]

He also urged that agricultural mechanization be based on the local collective economy and on self-reliance. This would mean that communes and teams would purchase most of the machinery themselves.

Hsiang recognized that these two requirements necessitated a partial reorganization of the agricultural machinery industry.

> First . . . the agricultural machinery mnufacturing departments . . . must more quickly shift the development of farm machines to the direction of taking small farm machines as the main target. Energetic efforts should be made to develop small horsepower power-operated machines, small threshing machines, rice and flour processing machines and machines for plant protection.

More than a change in design and priorities was required, however. A basic shift in the organization of the farm machinery factories was needed:

> the production of farm machines should mainly depend on local medium-sized and small enterprises for adopting methods of specialized production and cooperation. These enterprises are required to increase the variety, improve the quality, lower the costs, and make every farm machine measure up to the requirements for good performance, high efficiency, low costs, and a long lifespan.[15]

As a practical matter, the mechanization Hsiang called for could not come immediately. He thus strongly recommended a transitional period of semimechanization, just as he had in his December 1962 article following the Tenth Plenum.[16] His summary of the logic of semimechanization is worthy of examination:

> 1. The peasants urgently demand changing the production conditions; . . . they are not willing to remain for long on the foundation of manual labor. . . .

14. *Ibid.*, p. 14.
15. Hsiang, "Agricultural Mechanization Can Be Achieved," SCMP 3518, p. 5.
16. Previously, Hsiang had argued historically, citing the growth of semimechanization in the United States, the Soviet Union, and Japan. Now, he argued empirically, based on China's needs; perhaps his investigation trip to the countryside sensitized him to new factors.

Yet the state is not able to supply large quantities of farm machines overnight. Hence the demand for semi-mechanized farm tools. Semi-mechanized farm tools are more efficient than old-fashioned small farm tools, and at the same time can be supplied by the state in large quantities.

2. [Semi-mechanized farm tools] are more suited to the present purchasing power of the rural people's communes and to the demand of most production teams which want to buy and use these tools themselves.

3. They can be operated conveniently and simply and can be repaired easily; therefore they are suited to the technical level of the countryside at the present.

4. They can be produced and used locally and are more suited to the local needs of agricultural production, particularly the needs for harnessing mountains and rivers.

5. Their application makes it possible to mobilize to the broadest extent the peasants to participate in the farm tool innovation movement and in the work of experimenting on manufacturing and popularizing semi-mechanized farm tools.[17]

An evaluation of mechanization prepared by T'ao Chu, Party leader in Kwangtung, reached similar conclusions. T'ao led a work team of the Central-South Bureau of the CCP Central Committee on an inspection tour of Kwangtung in the spring of 1965. The work team observed that "water conservation construction in most of the areas of Kwangtung can be said to have been basically completed."[18] They paid special attention to whether the new conservancy program was producing any labor surpluses, and reported:

At one time . . . production activities of basic-level production teams were unplanned and there was the problem of superfluous labor power. To employ their surplus manpower, some production teams were willing to restore the use of the old-type water wheels for drainage and irrigation. They were not willing to use the newly-built electric-operated drainage and irrigation equipment. This was a new problem, and a big problem at that, which Chouhsin commune had to solve urgently if it were to continue its onward march.

The commune solved these problems by diversifying its economy. Industrial crops (including peanuts, onions, garlic, sugarcane, and bamboo) were sown, more livestock was raised, and local industry—a brick kiln—was developed. Leadership from a Party committee at a higher level was required to bring about this diversification.[19]

17. Hsiang, "Agricultural Mechanization Can Be Achieved," SCMP 3518, p. 6.

18. Work Team of Central-South Bureau of the CCP Central Committee, "Where Should Revolutionary Zeal be Exerted? Second Question Concerning Organization of a High Tide in Agricultural Production," *Nan-fang Jih-pao*, March 28, 1965 (SCMP 3437, p. 12).

19. *Ibid.*, p. 10.

Table 9.2. Agricultural diversification, Talang Commune, Tungkuan County, Kwangtung

Year	Gross farm output (yuan)	% grain	% diversified	Cash distributed (yuan)	Communal accumulation (yuan)
1957	2,833,264	55	45	1,414,542	107,654
1962	5,618,617	36	64	3,146,364	377,857
1963	5,854,852	35.7	64.3	3,192,740	427,095
1964	6,215,597	38	62	3,231,200	621,127

Source: Wang Chen-hua, "Diversified Economic Undertakings Promote Development of Grain Production," *Jen-min Jih-pao,* Oct. 21, 1965 (SCMP 3577, p. 14).

Other examples of what mechanized irrigation could do in South China were reported in the Chinese press. In 1963 a commune in Tungkuan County, Kwangtung, about 40 km. east of Canton, got electric machinery for irrigation. With released labor and land, the commune was able to diversify its agriculture, with peanuts, bamboo, and fruit, expand pig raising, and develop small industries such as farm implements, charcoal, and brick and tile. As diversification proceeded, both individual cash income and communal accumulation increased (see Table 9.2).[20] Outside Wuhan, the impact of mechanization was to raise the cropping index. Three-fourths of the rice fields switched from one to two crops of rice. Also, cultivation was intensified.[21]

Intermediate Technology, 1964–66

Agricultural mechanization policy had two distinct threads in the years just before the Cultural Revolution. There were simultaneous campaigns to consolidate and focus tractorization in a few specific localities (described in the previous chapter) and campaigns to encourage nationwide semimechanization. As the struggle within the top leadership intensified these two policies came more into conflict. Rather than "walking on two legs," they began to "struggle between two lines."

From 1964 to 1966, soon after Hsiang Nan's trip, there was a new

20. Wang Chen-hua, "Diversified Economic Undertakings Promote Development of Grain Production—Report of an Investigation into the Development of Diversified Economic Undertakings in Talang Commune, Tungkuan Hsien, Kwangtung," *Jen-min Jih-pao,* Oct. 21, 1965 (SCMP 3577, p. 14).

21. "Communes Equip Themselves with Machinery, Raise Farming Level," NCNA Wuhan, Dec. 4, 1965 (SCMP 3593, p. 20).

campaign to encourage semimechanization.[22] A directive concerning semimechanization probably came out in 1964; although the text is not available, the directive apparently included guidelines for the development of local factories to stress semimechanized farm implements.[23] Teng Chieh, vice minister of the Second Ministry of Light Industry (which was established in early 1965 to coordinate the new semimechanization campaign) indicated the general thrust of the new policy: "In 1964 the CCP Central Committee and the State Council proposed the 'policy of developing mechanization and semi-mechanization simultaneously with emphasis to be laid on semi-mechanization for a long time to come.' "[24]

In conjunction with this new directive the Ministry of Agricultural Machinery and the Ministry of Agriculture sponsored a national conference on semimechanized farm equipment in October 1964. The conference recalled the great successes of the tool improvement campaign of 1958–59, which had been ridiculed by some. A *People's Daily* editorial published coincidentally with the conference recalled Mao's statement in 1959, "Mechanization is the way out for agriculture," and cited the directive from 1959 calling for the establishment of locally based farm-tool study centers and tool improvement.[25]

While the conference looked back with a certain nostalgia on the 1958–59 campaigns, it took great care to emphasize that the errors should not be repeated. Tools must be carefully tested to fit the needs of each locality: "For the sake of meeting the requirement of the high-tide in agricultural production, it is necessary to intensify popularization of semi-mechanized farm equipment and implements, to persevere in scientific experiments, the mass line, and the principle of adopting means apro-

22. Not surprisingly, the new efforts coincided with broader rural policy. In June 1964 and December 1964–January 1965 the Party leadership held central work conferences to discuss rural policy. In both, Mao put forward proposals which were later distributed as directives. From the June 1964 meeting came the organizational rules for Poor and Lower Middle Peasant Associations; from the later meeting came "the Twenty-Three Articles," dealing with problems arising from the rural socialist education movement. See Richard Baum and Frederick Teiwes, *Ssu-Ch'ing, The Socialist Education Movement of 1962–1966* (Berkeley: University of California Center for Chinese Studies, 1968).

23. Presumably regional bureaus sent out similar and more detailed directives, and there is specific reference to one from the North China Bureau in Yüeh Tsung-t'ai, "Local Machine-Building Industry Should Wholeheartedly Serve Agriculture," *Jen-min Jih-pao,* Aug. 18, 1965 (SCMP 3531, p. 2).

24. Teng Chieh, "Bring the Role of Handicrafts into Full Play, Serve Agricultural Production Better," *Jen-min Jih-pao,* Jan. 4, 1966 (SCMP 3619, p. 7).

25. "The Spirit of Self-Reliance Must Be Upheld in Farm Implement Innovation Movement," *Jen-min Jih-pao* editorial, Oct. 15, 1964 (SCMP 3329, p. 14).

priate to local conditions and reality, and to innovate farm tools in a down-to-earth manner."[26] The errors of improperly distributing double-bladed plows, cable-drawn plows, and rice transplanters were thus implicitly recognized and criticized.

One initial response to this conference and the editorials praising it seems to have been puzzlement. The readers of *People's Daily* sent in questions about the editorial. The *People's Daily* editor summed up the inquiries this way: "The readers are of the opinion that while it is relatively easy to understand 'mechanization,' they are at a loss to know about 'semi-mechanization.' What are its limits? How is it to be distinguished from improvements of farm tools? Which are semi-mechanized machines and tools? They are not clear about these questions."[27] This puzzlement may have been, in reality, lack of enthusiasm for or opposition to the new semimechanization campaign.

Organizational changes within the Chinese government at about this time seem to have highlighted the new emphasis on semimechanization. The Ministry of Agricultural Machinery was renamed the Eighth Ministry of Machine Building on January 4, 1965. (This change was coincident with the Central Work Conference of December 1964–January 1965 that put out the Twenty-three Articles.) Although the top personnel of the ministry were unchanged, the change in name may have indicated that the ministry was specializing in advanced machinery and was in fact more closely related to industry than to agriculture.

A new ministry, the Second Ministry of Light Industry, was created on February 20, 1965, to the campaign to develop semimechanization and improved tools. In January 1966 one of its vice ministers, Teng Chieh, wrote an article in *People's Daily* entitled "Bring the Role of Handicrafts into Full Play, Serve Agricultural Production Better,"[28] which stressed the role of handicrafts industries in supplying semimechanized farm implements.

It is surprising that there was no organization to coordinate production of improved and semimechanized farm tools before 1965. Perhaps the new Second Ministry of Light Industry in some sense replaced the Na-

26. "National Conference on Semi-Mechanized Farm Equipment and Implements Decides on Conducting of Movement of Innovating Farm Equipment and Implements," *Jen-min Jih-pao,* Oct. 15, 1964 (SCMP 3329, p. 14).

27. "What are Semi-Mechanized Farm Machine and Tools?" *Jen-min Jih-pao,* Nov. 9, 1964 (SCMP 3342, p. 13).

28. Translation available at SCMP 3619, p. 6.

Table 9.3. Reports of new tools in 1965

Province	Report	Date
Liaoning[a]	50 new tools in past five months	June 26, 1965
Shansi[b,c]	400 new implements, farm tool study group	May 10, 1965
Shantung[d]	63,000 pieces in six weeks	Feb. 20, 1965
Inner Mongolia[c,e]	Successful innovations reported	Oct. 11, 1965
Hunan[c]	Farm tool renovation team	Jan. 1965
Hainan Island[f]	70 kinds of tools	Aug. 19, 1965

Sources:

a. "More New Farm Machinery Produced in Northeast China," NCNA Shenyang, June 26, 1965 (SCMP 3489, p. 15).

b. "North China Exhibition of Improved Farm Implements," NCNA Taiyuan, May 10, 1965 (SCMP 3457, p. 23).

c. "News about Agricultural Machinery," *Chung-kuo Nung-yeh Chi-hsieh,* no. 1 (Jan.), 1965, pp. 2–3 (JPRS 42,484, p. 1).

d. "East China Province Makes More Farm Machinery and Implements," NCNA Tsinan, Feb. 20, 1965 (SCMP 3404, p. 18).

e. "Chinese Farm Machinery Plant Wins Peasants' Applause," NCNA Huhehot, Oct. 11, 1965 (SCMP 3558, p. 22).

f. "Hainan Island Makes Farm Machinery," NCNA Canton, Aug. 19, 1965 (SCMP 3524, p. 16).

tional Office for Farm Tool Reform, which had been discontinued, according to Red Guards.

While statistics for 1965 are lacking, the press carried many articles indicating that a large number of new tools were locally designed and produced that year. In March 1965 it was claimed that 200 new types of farm machines had been developed in the previous year and that many were in mass production.[29] By October, a National Farm Tools Exhibition in Peking (organized by the Second Ministry of Light Industry in conjunction with the Ministry of Agriculture and the All-China Federation of Handicraft Cooperatives) showed 500 new varieties of semi-mechanized and improved farm tools.[30] From various provinces came similar reports (see Table 9.3).

During this period, research continued on two important tools for paddy cultivation: the cable-drawn plow developed in 1958 and worked

29. "China Produces More New Farm Machinery," NCNA Peking, March 16, 1965 (SCMP 3420, p. 26).

30. "National Farm Tools Exhibition in Peking," NCNA Peking, Oct. 26, 1965 (SCMP 3569, p. 19).

on through 1963,[31] and the rice transplanter. In 1966, the Kiangsu Provincial Agricultural Machinery Company had produced 1,600 cable-drawn plows for testing. A major improvement was in the durability of the cable.[32] An improved rice transplanter was tested by the Nanking Agricultural Mechanization Research Institute of the Chinese Academy of Agricultural Sciences in January 1965. It was smaller and slower than the earlier model, transplanting 12 mou a day.[33] By the end of the year, a factory had been set up in Nanning, Kwangsi, to produce another new model, the Kwangsi-65 rice transplanter. It sowed five rows and the number of shoots as well as the depth of planting could be adjusted.[34] In addition to the models previously tested, there were by 1965 a few other new versions: the Shanghai-65, the Hupeh-59, and the Fengyang-6016 power-driven transplanter.[35]

The agricultural tool works in the Fuhsin Mongolian Autonomous County in Liaoning was publicized for emulation at this time. This factory was commended for close work with the surrounding area, for investigation and study, for its attention to "on the spot" service and repair, and for its sensitivity to agricultural schedules.[36] Another model was a commune which developed 22 new implements and fabricated 12,000 items.[37]

The Arguments against the Trusts

Mao and some other Chinese leaders clearly disliked the plan to consolidate and expand the trusts to control all aspects of agricultural mechanization (and other sectors of the economy as well). Reasons for this were diverse.

31. "Technical Transformation of China's Agriculture," NCNA Peking, Sept. 30, 1963 (SCMP 3074, p. 7).

32. Shih P'ing, "Some Suggestions Concerning the Use and Study of Electrical Appliances for Farming Purposes," *Chi-hsieh Kung-yeh,* no. 3, 1963 (SCMM 362, p. 21); Sun I-k'un (Kiangsu Provincial Agricultural Machinery Company), "Points of Attention to be Paid in Using Cable-drawn Plows," *Nung-yeh Chi-hsieh Chi-shu,* no. 9 (Sept.), 1966, p. 30 (JPRS 39,619, p. 9).

33. "New Machines for South China Paddy Fields," NCNA Nanking, Jan. 10, 1965 (SCMP 3377, p. 18).

34. "China's First Rice-Transplanter Factory Goes into Production," NCNA Nanning, Nov. 14, 1965 (SCMP 3581, p. 15).

35. "Field Experiment of Water Paddies Sprout Planters," *Chung-kuo Nung-yeh Chi-hsieh,* no. 7 (July) 1965, cover 2 (JPRS 42,484, p. 52).

36. "Yüeh, "Local Machine-Building Industry," SCMP 3531, p. 2.

37. "Famous Yangtze River Island Commune Designs Tools for Cotton Cultivation," NCNA Nanchang, June 11, 1965 (SCMP 3478, p. 14).

The simple factors can be examined first. One reason the trust system was rejected is that it required manufacture and supply of certain specialized parts of engines to become increasingly centralized. This was expected to increase production efficiency. However, the centralized production of crucial parts required efficient transportation to other factories. Close coordination and careful timing were essential to permit all factories to operate at capacity without excessive inventories. Substantial problems emerged in making this transition. For example, an engine factory in Shanghai previously produced its own oil transmission pumps. The trust centralized production of the pumps in another factory in Tientsin; but the new manufacturer could not satisfy the needs of the factory in Shanghai. In addition, other factories which used pumps supplied by the Shanghai factory could not obtain them from Tientsin, and so had to cease production of engines. Finally, the Shanghai plant resumed production of the pump.[38] A similar problem was caused when copper washers used by factories in Szechwan were to be supplied by factories in Anhwei.[39] Of course such problems of coordination must be expected in the course of a major change in patterns of supply to factories, and it must be presumed that China could have ironed out such problems after a short while.

It is, however, very possible that some Chinese leaders—especially Mao—did not desire developing an economy which required sophisticated integration and coordination. Such an economy would be particularly vulnerable to serious disruption by natural disaster or military attack. The military implications of such an economic system may have been particularly important during this period of 1965 and 1966 when the U.S. air attacks and troop invasions on the Asian mainland were escalating rapidly. Even if Mao downgraded the risk of war with the United States, he was not likely to ignore the military situation.

The trust system of organization was also attacked by county- and province-level officials, who would be deprived of any influence over the mechanization of agriculture by the trust system. They would lose the ability to subsidize economically poor production teams or politically advanced teams. They would lose their influence on hiring practices of the

38. Peking Agricultural Machinery College, East Is Red Commune, Criticism and Repudiation Office, "Completely Settle the Heinous Crimes of China's Khrushchev and Company in Undermining Agricultural Mechanization," *Nung-yeh Chi-hsieh Chi-shu*, no. 5 (Aug. 8), 1967 (SCMM 610, p. 27).

39. *Ibid.*

agricultural machinery system. In addition, the 100 keypoint county plan meant that roughly 95 per cent of China's counties would not receive state aid in mechanization for several years. (Probably more than 95 per cent of China's population would not see agricultural machinery, as it would be concentrated in sparsely populated regions.) For these reasons, the plan was opposed by many provincial representatives, according to Red Guards.[40]

Another problem was financing the trust. By relying on state investment, the opportunities to develop local enthusiasm, to raise money locally, to train local personnel, and to encourage local responsibility for operation and maintenance of machinery would be lost. Mao's feelings about the importance of these factors with regard to agricultural machinery had been set forth in his statement at Chengtu in March 1958 and were repeated in his letter of March 12, 1966.[41]

Finally, the plan seemed to envision a process of many decades, because it would rely on profits from early investments in machinery to finance future growth. Mao had always been interested in fairly rapid mechanization throughout China because he viewed mechanization as contributing to the consolidation of the collective economy and to a lessening of the "three differences" (the differences between urban and rural life, between industrial and agricultural work, and between intellectual and manual work). He may have rejected the enlarged agricultural machinery trust because it would not bring about rapid mechanization on a widespread basis.

It is interesting that there was no criticism of either the trust plan or the 100 keypoint county plan on the grounds that these plans would increase inequalities of income between communes. Whether the plans would have increased inequalities would have depended on price and tax policy. Perhaps there was no such criticism because both the leaders and the Red Guards realized that income inequalities would not be eliminated—and might even be increased—with the policy of self-reliance, because only those communes with capital and favorable natural conditions would purchase and profit from agricultural machinery. Indeed, the only mention of increasing income inequality was in conjunction with criticizing a

40. Eighth Ministry of Machine Building, Agricultural Machinery Management Bureau, Revolutionary Great Alliance Committee, Mass Criticism Unit, "T'an Chen-lin's Crime of Sabotage against Farm Mechanization Must Be Reckoned With," *Nung-yeh Chi-hsieh Chi-shu,* no. 6 (June 8), 1968 (SCMM 624, p. 7).

41. See p. 199 above and p. 224 below.

proposal that seems to be rather close to the Maoist idea of self-reliance. T'an Chen-lin reportedly suggested: "If communes are to own the machines, we should sell them to anyone who comes up with the money. . . . Satisfy their demands and give them the best machines."[42] Red Guards criticized this suggestion with this argument: "If things were allowed to go on in this way, the gaps between the rich and poor communes would be inevitably widened and socialist direction would be ignored."[43]

The main problem with the trusts seems to be that they were inspired to some extent by Soviet forms of economic organization, which Mao perceived as revisionist. Adoption and expansion of this form of organization would inevitably mean that China would be following the Soviet model.[44] In September 1965, after years of discussion, the Soviet Union announced a plan for economic reorganization, commonly called Libermanism. On general principles, one would expect many Chinese to have shown interest in these reforms. The Soviet Union had, after all, been enshrined as the model for socialist development during the early 1950's, and undoubtedly many Chinese continued to look there for ideas about building socialism and modernization. Moreover, the Liberman reforms appeared to introduce a certain degree of flexibility and economic pragmatism into the economic planning system, and this may have been as welcome to some in China as to those in some Eastern European countries.

Liu Shao-ch'i, apparently, was impressed by Libermanism, and wanted to introduce more freedom for enterprises to maximize profits and to hire and fire workers without centralized controls.[45] In 1964, K'ang Sheng warned Mao: "Sun Yeh-fang of the Economic Institute has been fooling around with Liberman's works and with capitalism."[46] Mao's initial reaction was remarkably lighthearted:

It is all right to engage in some capitalism. Society being so complex, wouldn't it be too monotonous to engage only in socialism to the exclusion of capitalism? Wouldn't that be too one-sided an approach, without any units of opposites? Let

42. "T'an Chen-lin's Crime of Sabotage," SCMM 624, p. 6. 43. *Ibid.*

44. In addition to Yugoslavia, perhaps Czechoslovakia (under Dubček) was also a model of development Mao feared as revisionist. Lowell Dittmer, *Liu Shao-ch'i and the Chinese Cultural Revolution* (Berkeley: University of California Press, 1974), p. 246.

45. Red Guard sources are cited by Christopher Howe, "Labour Organization and Incentives in Industry," in Stuart Schram, ed., *Authority, Participation and Cultural Change in China* (Cambridge: Cambridge University Press, 1973), p. 242.

46. Talk on Problems of Philosophy, Aug. 18, 1964, *Miscellany*, p. 386.

them engage in it. Let them attack us madly, demonstrate in the streets, take up arms to rebel—I approve all of these things. Society is very complex, there is not a single commune, a single county, a single department of the Central Committee, in which one cannot divide into two.[47]

While Mao often tried to structure a situation in which conflict would uncover the mistakes and insensitivities of middle-level cadres, it is more likely that he lacked the political support in 1964 to stop discussions and experiments with Liberman's works. By 1966, however, he apparently thought choices would have to be made; the tension within the unity of opposites was too great. A *People's Daily* editorial on April 3, 1966, at the time of the discussion about trusts, posed the question sharply: "In regard to production and technique, management, regulations in our enterprises, shall we go our own way or copy from capitalism and revisionism?"[48] This same editorial suggested that the broadest social and ideological factors were involved in the selection of an industrial management system:

Shall we foster the concept of serving the customer, that is, serving the people, whole-heartedly, or shall we adhere to the idea of considering only fulfilling production quota and gaining profit? Shall we work hard and adhere to the principle of running enterprises with industry and thrift, or shall we indulge in an easy and comfortable life and discard that principle? Shall we foster a collective spirit and a communist style, or cultivate borgeois ideas? Shall we gradually narrow down the differences between town and country, workers and peasants, and mental and physical labor or preserve and widen them?[49]

By 1964, Mao had already reached the judgment that the Soviet Union had become a revisionist country. In May he said: "The Soviet Union today is a dictatorship of the bourgeoisie, a dictatorship of the grand bourgeoisie, a fascist German dictatorship, and a Hitlerite dictatorship. They are a bunch of rascals worse than De Gaulle."[50] The next month he observed: "In the Soviet Union, it was the third generation that produced the Soviet Khrushchev Revisionism."[51] Mao became increasingly con-

47. *Ibid.* Also available in Stuart Schram, ed., *Chairman Mao Talks to the People* (New York: Pantheon, 1974), p. 216.

48. "Running Enterprises in Line with Mao Tse-tung's Thinking," *Jen-min Jih-pao* editorial, April 3, 1966; available in *Peking Review,* no. 16 (April 15), 1966, p. 11. Emphasis added.

49. *Ibid.*

50. Some Interjections at a Briefing of the State Planning Commission Leading Group, May 11, 1964, *Miscellany,* p. 349.

51. Talk on Putting Military Affairs Work into Full Effect and Cultivating Successors to the Revolution, June 16, 1964, *ibid.,* p. 357.

cerned with the task of making sure that China did not follow this Soviet path: "We can also possibly produce revisionism. How can we guard against revisionism? How can we cultivate successors to the revolution?"[52] "If one does not pay attention, there is bound to be revisionism."[53]

Mao described the challenge of revisionism in a remarkable discussion with Andŕe Malraux on August 3, 1965.

> There are now two paths for every communist: that of socialist construction, and that of revisionism. . . . Corruption, law-breaking, the arrogance of intellectuals, the wish to do honor to one's family by becoming a white-collar worker and not dirtying one's hands anymore, all these stupidities are only symptoms. Inside the party and out. The cause of them is the historical conditions themselves. But also the political conditions. . . . Humanity left to its own devices . . . does reestablish inequality. The forces tending toward the creation of new classes are powerful. . . . But in this battle [against revisionism] we are alone. . . . I am alone with the masses. Waiting. . . . Revisionism is the death of the revolution.[54]

What, exactly, were the specific elements of revisionism that worried Mao so much, and how were they manifested in agricultural mechanization? A systematic critique of the Soviet Union's policies, *On Khrushchev's Phoney Communism and Its Historical Lessons for the World* (often attributed to Mao himself), was issued by the Chinese leadership on July 14, 1964.[55] This document charged that leading functionaries of state-owned factories amass large fortunes; that leading collective-farm functionaries and their gangs steal and speculate at will, freely squander public money; that people set up private enterprises, and engage in commercial speculation; that a privileged stratum has emerged of highly paid factory and farm managers and bourgeois intellectuals; that material incentives have been adopted, and income distribution has become less equal; that the principle of profit and competition are undermining socialist ownership and planning; and that bourgeois culture, including philosophy and values, are spreading in the Soviet Union.

The trusts advocated (and experimentally established) in China were,

52. *Ibid.*

53. Interjections at an Anti-Revisionist Reports Meeting, Sept. 4, 1964, *Miscellany,* p. 407.

54. André Malraux, *Anti-memoirs* (New York: Holt, Rinehart, & Winston, 1967), pp. 369, 370, 373, 375.

55. Editorial Departments of *People's Daily* and *Red Flag, On Khrushchev's Phoney Communism and Its Historical Lessons for the World* (Peking: Foreign Languages Press, 1964).

unquestionably, not precise copies either of the previous Soviet trusts or of the Liberman reforms. The Liberman reforms brought economic accounting and profit goals to each factory, while the Chinese trusts put economic accounting on an industrywide basis. The distinction should not be pressed too hard, however, as the Chinese plans for trusts seemed to include enterprise-level accounting, and the Soviet reforms included strengthening "cost-accounting amalgamations," i.e., trusts. On a broader level, the Chinese trusts may very well have been inspired by some of the ideas which formed the foundation of the Liberman reforms, namely, the use of profit as a criteria for economic efficiency and the use of material incentives, particularly bonuses for managers and piece-rate payment to workers.[56]

Having determined to his own satisfaction that the Soviet Union was already revisionist, Mao saw the Liberman reforms as a way of consolidating the new capitalist system there. In fact, the Chinese charged that the Liberman reforms represented a conscious effort to integrate Western capitalist managerial experiences with Soviet practice:

> In November 1962 Khrushchev said at the plenary session of the Central Committee of the Soviet Communist Party: "The Soviet Union should learn as much as possible from the advanced West."
>
> [In the early 1960's] the Soviet modern revisionist clique had on numerous occasions sent people to America and other capitalist countries to study their "experience" in enterprise management, apart from inviting British managers and advisers to give talks on the "experience" in Western enterprise management. At the same time, no efforts were spared in the Soviet Union to push and carry out Liberman's propositions of "planning, profits and cash awards," managing enterprises completely in accordance with the principle of putting the ruble in command.[57]

56. Liberman's own description of the system is conveniently available in E. G. Liberman, *Economic Methods and the Effectiveness of Production* (New York: Anchor, 1973), esp. pp. 11–49. A group of Soviet economists has also described the system; *Soviet Economic Reform; Progress and Problems* (Moscow: Progress Publishers, 1972). An excellent description of the way bureaucratic controls in the Soviet Union discouraged technological innovation in most sectors of civilian industry, construction, commerce, and services is available in Gertrude Schroeder, "Soviet Technology: System vs. Progress," *Problems of Communism* 19:5 (Sept.–Oct. 1970), 19–30. Difficulties in interenterprise coordination of materials and equipment, short-term production targets, and irrational prices are stressed. The Chinese stress on self-reliance within an enterprise or region is clearly a way of reducing the inefficiencies and problems of coordination inherent in the Soviet pattern of industrial organization.

57. Ching Hung, "The Plot of the Top Ambitionist to Operate 'Trusts' on a Large Scale Must Be Thoroughly Exposed," *Kuang-ming Jih-pao,* May 9, 1967 (SCMP 3948, p. 6). It

The Chinese have charged that the Liberman reforms were successful in consolidating revisionism in the Soviet Union. In 1967 China's leadership argued:

> Since it came into power, the Soviet revisionist leading group has energetically enforced "reforms of the economic system" in the different branches of the national economy. These were mainly based on Liberman's recommendations. . . . The essence of the "new system" being pushed ahead so vigorously by the Soviet revisionist group under the cloak of "economic reform" is to practise in an all-round way capitalist management in all fields of the national economy, completely disrupt the socialist relations of production and thoroughly break up the socialist economic base.[58]

In retrospect, Mao's concerns seem to have had some basis. In the years since the early 1960's, there is substantial evidence that an elite has become more differentiated in the political, economic, and intellectual sectors of Soviet society, and that this elite is becoming increasingly wealthy.[59] At the same time, workers are more subject to unemployment, and feelings of alienation are widespread.[60] During this period the overall growth rate of industry was, if anything, a little lower, dropping from 9.8 per cent in the 1950's to 5–7 per cent in later years.[61] However, the con-

---

might be noted that Mao endorsed the notion of learning from the United States in another context: "Mao said that China should learn from the way America developed, by decentralizing and spreading responsibility and wealth among the fifty states. A central government could not do everything" (Edgar Snow, *The Long Revolution* [New York: Vintage, 1973], p. 175).

58. *How the Soviet Revisionists Carry out All-Round Restoration of Capitalism in the U.S.S.R.* (Peking: Foreign Languages Press, 1968), pp. 2–3. This pamphlet is a collection of articles from late 1967. This charge is restated in the important polemic "Leninism or Social Imperialism?" *Jen-min Jih-pao* editorial, April 22, 1970; available in *Peking Review*, no. 17 (April 24), 1970, p. 8.

59. This is the perception of many who left the Soviet Union, and it agrees with surveys made by Soviet social scientists. Zev Katz, "Insights from emigres and Sociological Studies on the Soviet Union," in *Soviet Economic Prospects for the Seventies* (Washington, D.C.: U.S. Congress Joint Economic Committee, June 27, 1973), pp. 87–121. Observations by Soviet economic planners agree. For example, one economist observed: "Inevitably there is a rise in the share of those groups that have relatively high incomes. At the same time there is a reduction in the share of families for which a comparatively low savings norm is characteristic." T. Ivensen, "Problems in Forecasting the Monetary Savings of the Population," *Ekonomicheskie Nauki,* no. 11, 1973; available in *Problems of Economics,* June 1974, p. 66, cited by Martin Nicolaus, in *Restoration of Capitalism in the USSR* (Chicago: Liberator Press, 1975).

60. Nicolaus, *Restoration of Capitalism,* pp. 107–114; Katz, "Insights," pp. 114–115.

61. Rush V. Greenslade and Wade E. Robertson, "Industrial Production in the U.S.S.R.," *Soviet Economic Prospects for the Seventies,* p. 271.

nections between these tendencies and the Liberman reforms are vague, especially because many observers believe the reforms were not implemented thoroughly in practice.[62] Regardless of how the Liberman reforms were implemented in the USSR, they constituted a model of organization abhorrent to Mao. He and some other leaders did not want to let their economy function autonomously from government leadership and be steered by profits and material incentives. They feared that the political system would lose the power to direct the benefits of modernization to particular sectors, that a new elite could emerge and become a new ruling class, that the character of personal and work relations would be changed in deleterious ways. They chose to "strengthen the dictatorship of the proletariat" and "put politics in command."

62. Liberman himself reported that by 1969 the plan had been put into effect in 32,000 enterprises producing more than 77 per cent of all industrial output (*Economic Methods,* pp. 24–34). However, Western observers are divided over the extent to which the reforms were put into effect. Nicolaus, in *Restoration of Capitalism,* suggests that the reforms were a watershed in Soviet history, marking a sharp break from previous patterns of socialist management to a revisionist system. Specifically, he argues that planning machinery has failed to plan the economy (p. 136); that enterprise managers were given virtually total freedom in economic decision-making (pp. 125–130); that market forces determine prices (pp. 136–138); that profit-seeking state banks shape the Soviet economy (pp. 143–149); and that social services, including housing, medical care, and child care, have been reduced (pp. 176–178).

In almost complete contradiction, Gertrude Schroeder argues that the reforms of 1965 have been "much attenuated," and have been "withering away . . . step by step, until [they] now virtually melt into the familiar, still flourishing landscape of the centralized Soviet command system" ("Soviet Economic Reform at an Impasse," *Problems of Communism* 20:4, [July–Aug. 1971], 36). She argues that in precisely those areas in which Nicolaus sees great changes, the Soviet economic bureaucracy has passed more and more regulations so that the reforms have no major effect. She argues that establishment of trusts has been limited (p. 39); that price control remains (pp. 42–43); that enterprise managers still have little discretionary power (p. 43); and that long-term planning is being strengthened ("Recent Developments in Soviet Planning and Incentives," in *Soviet Economic Prospects for the Seventies,* pp. 11–38).

On two points there seems some agreement: that there are pressures for laying off workers; and that social services have been given less priority. (For Schroeder on this point, see "Consumer Problems and Prospects," *Problems of Communism* 22:2 [March–April 1973], 19.) Considering the widespread divergence in conclusions, it is clear that continued analysis of this important question will be needed.

The relationship between the reforms and economic growth rate is even more obscure. Nicolaus implies that a drop in growth rate is the natural result of the Soviet Union's following the capitalist road. Schroeder argues precisely the reverse: that the decline has come about because the reforms were not implemented thoroughly and economic affairs are still controlled by bureaucrats and are not capitalist enough. While Schroeder and Nicolaus are at opposite ends of a political and analytical spectrum, they both agree that the reforms implemented in the Soviet Union after 1965 did little to solve economic problems and may have contributed to them.

Carrying Out the Cultural Revolution

On the eve of the Cultural Revolution, sharp conflict emerged within the highest levels of leadership over alternative strategies of agricultural mechanization. Of course, mechanization was not the sole crucial issue leading to the Cultural Revolution; but it may have been one of the final straws that broke the proverbial camel's back. Agricultural mechanization was one of the last substantive issues discussed before the Cultural Revolution developed its own dynamics. The debate involved a head-on clash between Mao and Liu Shao-ch'i; it also involved P'eng Chen, just a few days before he was purged.

The struggle flared into the open in January 1966. The CCP Hupeh provincial committee decided to speed up dramatically its agricultural mechanization program in view of the opening of two tractor factories in Wuhan, making 10,000 7-h.p. walking tractors and 20-h.p. tractors a year. It was assumed that 22,000 of the 38,000 production brigades in the province could use two tractors each, and thus agriculture could be mechanized basically in five years.[63]

According to Red Guards, on February 19, 1966, Mao Tse-tung approved the plan, although he seems to have considered it a bit too rapid. At the same time he criticized the laxness toward agricultural mechanization in previous years: "Now is the time for us to lay hold of the next 15 years. A decade has already elapsed but our grip on that decade is by no means satisfactory."

Four days later, on February 23, Liu Shao-ch'i reviewed Mao's comments on the Hupeh plan and expressed reservations. Liu noted that "such work implicates many aspects of work," and according to the Red Guard account, Liu "venomously forbade the immediate transmission of Chairman Mao's instruction and the Hupeh proposals to all places." Only the relevant departments of the Central Committee were informed of the Hupeh plan. Moreover, he asked some other "top capitalist roaders" to comment on it.[64]

63. The prime source documenting this aspect of policy debate is: China Institute of Research in Agricultural Mechanization, Capital Workers' Congress, East Is Red Commune, "Last-Ditch Struggle that Courts Self-Destruction—Denouncing the Towering Crimes of China's Khrushchev in Opposing Chairman Mao's Wise Decision," *Nung-yeh Chi-hsieh Chi-shu,* no. 5 (Aug. 8), 1967 (SCMM 613, pp. 24–28).

64. Perhaps the other aspects of work "implicated" were the question of how labor displaced by mechanization would be utilized, and the supplying of additional steel to Hupeh to make the tractors. In addition, Liu may have wondered whether some of the tractors produced in Wuhan should be distributed to other provinces.

On March 11, Liu Shao-ch'i wrote to Mao suggesting that "the transmission of the Hupeh Provincial Committee's documents to all places should be postponed" until investigators could be sent to Hupeh to check on the plan. The following day, in a letter (described above, p. 199) on agricultural mechanization, Mao countered with his own proposal, and urged that Hupeh be visited as a model rather than as a site for investigation: "Giving tit for tat, he advocated that the various bureaus of the Central Committee, and the Party committees of various provinces, municipalities and regions should also send men to Hupeh to make a joint study. . . . On their return, each should draft a five-year, seven-year, or ten-year plan."[65]

The first days of April appear to be critical and confused. Red Guard sources say that on April 2, Liu Shao-ch'i, sensing the seriousness of the Cultural Revolution, secretly ordered his associates to "play a more active role" and reach for a compromise that would not exclude them from influence. On that very day Chou En-lai agreed to join Mao's attack on P'eng Chen.[66] On the next day, April 3, P'eng received a secret order from Liu Shao-ch'i to prepare a draft for a Central Committee comment to accompany the Hupeh proposals.[67]

There is a problem with this chronology in that Liu traveled to Pakistan and Afghanistan between March 26 and April 8, and to Burma and Pakistan from April 15 to April 19. It is not clear how Liu could have sent instructions on April 2 if he was abroad.

As for the charge that P'eng Chen received instructions on April 3 (when Liu was abroad), possibly the instructions had been issued before Liu left and were not received by P'eng until April 3. Although P'eng was to be purged shortly, as late as March 26 he gave a major speech welcoming Kenji Miyamoto, of the Communist Party of Japan,[68] so he may still have had substantial power. The critical turning point for P'eng came a few days later at a conference from April 9 to 12, at which P'eng

65. "Last-Ditch Struggle," SCMM 613, p. 26. The text of Mao's letter of instructions is available in *Miscellany,* p. 373, and *Peking Review,* no. 52 (Dec. 26), 1977, pp. 7–9.

66. Perhaps the more important issue involving P'eng Chen at this time had to do with cultural matters. As head of a Cultural Revolution group investigating the political implications of theatrical plays which obliquely attacked Mao, P'eng tried to downgrade the questions as "academic," in a statement of February 7, 1966. This undoubtedly led Mao to the decision that P'eng had to be purged. Edward Rice, *Mao's Way* (Berkeley: University of California Press, 1974), pp. 233, 239.

67. "Last-Ditch Struggle," SCMM 613, p. 27.

68. "Rousing Welcome for Japanese Communist Party Delegation," *Peking Review,* no. 14 (April 1), 1966, pp. 7–8.

was criticized. This conference prepared the way for another meeting on April 16 that dissolved the Cultural Revolution Group, of which P'eng was chairman, and was the first phase of his purge.[69]

The draft that P'eng prepared concerning the Hupeh proposal was not considered satisfactory because it neglected to emphasize the role of provincial responsibility and self-reliance and failed to repeat Mao's criticisms of the Soviet policy toward mechanization and of centralized controls. This draft, presumably, was not released.

Finally, according to the Red Guard account, on April 10, 1966, the Hupeh proposals were cleared for release under peculiar circumstances. Liu Shao-ch'i's letter of February 23 and Chairman Mao's comment of February 19 were both included at the same time as documents of equal standing.

(The Red Guards fail to explain why the Hupeh proposals were published without editorial comment on April 4 in the Canton *Evening News*.[70] Could Mao have leaked the report early to break the logjam in Peking and force some sort of comment to appear quickly to put the article in perspective? Could T'ao Chu, Kwangtung Party leader who controlled the Canton media and was elevated in August 1966 to the fourth-ranking person in the Party, have ingratiated himself with Mao by assuring publicity for this article?)

The Hupeh proposals were carried in *People's Daily* on April 9, 1966, along with a curious "editor's note," which could very well have been a combination of Mao's comments and Liu's letter. The first four paragraphs emphasize the importance of agricultural mechanization in developing a socialist society:

> It is only with agricultural mechanization . . . that . . . the rural economic outlook can be fundamentally changed, the differences between the city and the countryside, between industry and agriculture, and between mental and physical labor can be gradually reduced, the worker-peasant alliance can be consolidated, restoration of capitalism can be prevented, and the socialist positions can be further consolidated and developed.[71]

69. Rice, *Mao's Way*, p. 239.

70. "Postulation by the CCP Hupeh Provincial Committee Concerning Gradual Realization of Agricultural Mechanization," *Yang-ch'eng Wan-pao*, Canton, April 4, 1966 (SCMP 3679, p. 8).

71. "Editor's Note on the Postulation by CCP Hupeh Provincial Committee Concerning Gradual Realization of Agricultural Mechanization," *Jen-min Jih-pao*, April 9, 1966 (SCMP 3679, p. 14).

This section also emphasized the importance of "placing politics first," of "opposing those practices which lead us to divorce ourselves from the masses and politics." It urged continuation of the socialist education movement.

In contrast, the final two paragraphs were mild. Although they made reference to Mao's fifteen-year timetable, they made no criticism of the past. Moreover, the last two paragraphs envisaged a small experimental program at a few key points, rather than a nationwide movement: "We should immediately act and take advantage of the next 15 years to gradually realize, on the basis of self-reliance, this great call of Comrade Mao Tse-tung experimenting with a few points to begin with and gradually multiplying them through uninterrupted efforts." [72] Such a final sentence was hardly an enthusiastic endorsement of the Hupeh proposal for mechanization in less than ten years.

The significance of the debate on the Hupeh proposals is highlighted by the fact that the articles from April 9 *People's Daily* were reprinted in *Agricultural Mechanization Technology,* the professional periodical for agricultural machinery, in its May 1966 issue.[73]

By April 1966, the various proposals for mechanization became subordinate to the broad political and power consideration of the Cultural Revolution. Within two months, P'eng Chen had lost his position, and Liu Shao-ch'i was under criticism. Whatever debates there had been over agricultural mechanization were compounded and supplemented by new debates concerning strategy and tactics of the Cultural Revolution.

Action was taken on the Hupeh proposals early in the Cultural Revolution. The Central Committee in July 1966 convened a national conference on agricultural mechanization in Hupeh with representatives from various provincial, municipal, and regional Party committees. There was no claim that Mao attended, but he was in Wuhan for his famous swim on July 16, 1966. (The precise date of the conference is not known.) The conference reportedly endorsed the policy of "mainly relying on the efforts of the collective economy, and of relying on local industries to manufacture farm machines and tools." [74] The Eleventh Plenum in August

72. *Ibid.,* p. 15.

73. "Gradual Realization of Agricultural Mechanization," *Nung-yeh Chi-hsieh Chi-shu,* no. 5 (May) 1966, pp. 2–4 (JPRS 36,976, p. 1).

74. Remarkably enough, this meeting was remembered ten years later when the Hupeh Province Party Committee developed another plan for agricultural mechanization. Radio Wuhan, May 13, 1976.

1966 "expressed complete approval of Chairman Mao's wise policy concerning the question of gradual materialization of plans and arrangements for agricultural mechanization."[75] Agricultural mechanization was one of nine broad policies decided then.[76] According to the Red Guard accounts, "The agricultural mechanization undertaking which had for a time been led astray by China's Khrushchev [Liu Shao-ch'i] thus returned to the course charted by Chairman Mao. Braving the winds and waves, it resumed its voyage along the correct course."[77]

Postcultural Revolution Policies

The broad guidelines for agricultural mechanization adopted during 1966 had these basic points:

1. Mechanization and tool reform would receive special priority in the process of agrotechnical reform. The 1970's would be the decade of farm mechanization.
2. Special attention would be given to semimechanization as a stopgap measure.
3. To a greater extent, farm tools would be manufactured locally.
4. Provinces and to some extent production units would be self-reliant, that is, they should raise money for machinery locally.
5. Machinery would be owned and managed by the production units, not by a state bureaucracy or an economic trust (with the exception of state farms).

Priority of Mechanization

These policies could not be implemented systematically, however, until revolutionary committees seized and consolidated power. By 1969 this had happened, and the new policy could be pursued. The October 1969 issue of *Red Flag* carried an article urging increased emphasis on agricultural mechanization. The article almost paraphrased Mao's directive of April 1959, ten years earlier, when it stated: "Every county must positively set up farm machine repairing and manufacturing plants, establish an industrial network for serving agriculture, and contribute

75. "Completely Settle the Heinous Crimes," SCMM 610, p. 21.

76. "Communique of the Eleventh Plenary Session of the Eighth Central Committee of the CCP," Aug. 12, 1966, SCMP 3762.

77. *Ibid*. Also see Liao Nung-ko, "Knock Down China's Khrushchev and Completely Discredit by Criticism His Line of State Monopoly," *Nung-yeh Chi-hsieh Chi-shu*, no. 2, 1968 (SCMM 620, p. 6).

greater strength to speeding up the technical reform of agriculture."[78] At the same time, in September and October 1969, conferences on agricultural mechanization were held in several provinces, including Kansu, Inner Mongolia, Anhwei, Shantung, Hunan, and Kwangtung.[79]

An article broadcast by Radio Shanghai on November 10, 1969, emphasized agricultural mechanization in very strong terms:

> Today, our worker-peasant alliance has entered the new stage of realizing agricultural mechanization. Agricultural cooperativization without agricultural mechanization cannot consolidate the worker-peasant alliance because it is impossible for such an alliance to rest forever on two diametrically opposed material and technical foundations.
>
> The development of agricultural production lies in mechanization. The realization of agricultural mechanization in turn will consolidate agricultural collectivization and eliminate the differences between workers and peasants, town and countryside, mental and manual labor.[80]

This argument indicated that peasants would consider themselves shortchanged and express discontent politically. It is interesting to note that in 1955, in "On the Question of Agricultural Cooperation," Mao had predicted that serious economic problems would emerge by 1970 if agricultural mechanization were not achieved:

> If we cannot fundamentally solve the problem of agricultural cooperation in a period of roughly three five-year plans [i.e., by 1970] that is to say, if our agriculture cannot make a leap from small-scale farming with animal-drawn farm implements to large-scale mechanized farming, including extensive state-organized land reclamation by settlers using machinery (the plan being to bring 400–500 million mou [27–33 million ha.] of waste land under cultivation in the course of three five-year plans), then we shall fail to resolve the contradiction between the ever-increasing need for marketable grain and industrial raw materials and the present generally low yield of staple crops, we shall run into formidable difficulties in our socialist industrialization and shall be unable to complete it.[81]

The Radio Shanghai broadcast revealed that opposition to emphasis on farm mechanization still existed:

78. Writing Group of Peking Municipal Revolutionary Committee, "The Road to China's Socialist Industrialization," *Hung-ch'i,* no. 10 (Sept. 30), 1969 (SCMM 666, p. 18). The article is available in *Peking Review,* no. 43 (Oct. 24), 1969, pp. 7–13, but for unknown reasons the passage quoted here is deleted.

79. Lin Chen, "The Agricultural Plight of the Chinese Communists," *Issues and Studies* 5:5 (Feb. 1970), 41–42.

80. Radio Shanghai, Nov. 10, 1969 (FBIS Nov. 19, 1969).

81. *Selected Readings,* p. 405.

> Some of our comrades maintain that supporting agriculture results in much trouble and yields little profit. They advance the theory of "indirectly supporting agriculture." This is extremely wrong. By seeking only profits instead of serving agricultural production, county and commune industries may superficially accumulate some funds, but, in fact, this only serves to promote capitalist tendencies.[82]

During 1970, there was an avalanche of news from all provinces describing progress in agricultural mechanization. An unpublished compilation done by British China watchers, made available on November 26, 1970, listed over 150 separate references to press and radio broadcasts, covering almost every province, describing advances in agricultural mechanization during that year. In general, the articles cited the development of local factories. A wide variety of tools was manufactured, from large tractors and diesel engines to the cable plows and rice transplanting machines in some southern provinces and threshing machines in Kwangtung. Local factories used raw materials, such as iron and coal, from within the province. Substantial emphasis was placed on servicing and repairing existing machines.[83]

The general structure of the agricultural machinery industry has been described in Chinese engineering books:

> In order to accelerate agricultural mechanization, a large number of middle and small size regional factories were constructed all over the country. Essentially every county has a tractor manufacturing plant or a tractor repair and assembly plant; in many regions there exists a repair net extending from the county level to the commune and the brigade levels.
>
> In addition, most regions are equipped with sizable tractor parts factories and middle to small size tractor manufacturing plants and diesel engine factories for farming purposes.[84]

Estimates by U.S. government analyst Robert Field, charted in Figure 9.1, indicate very rapid growth of the tractor industry in the post–Cultural Revolution years. Chinese economists had estimated that roughly one million standard (15-h.p.) tractors would be needed to tractorize China's agriculture (see Table 7.8). By 1973, according to computations based on

82. Radio Shanghai, Nov. 10, 1969 (FBIS Nov. 19, 1969).

83. A resurgence of policies reminiscent of the Great Leap during 1970 has been noted by Harry Harding, "Political Trends in China since the Cultural Revolution," American Academy of Political and Social Sciences, *Annuals* 402 (July 1972), 71.

84. *T'o-la-chi Kou-tsao* [Structure of Tractors] (Peking, 1973), available in *Translations on People's Republic of China,* no. 287 (JPRS 63, 091), p. 18.

Figure 9.1. Production and cumulative total of tractors in China, 1958–73

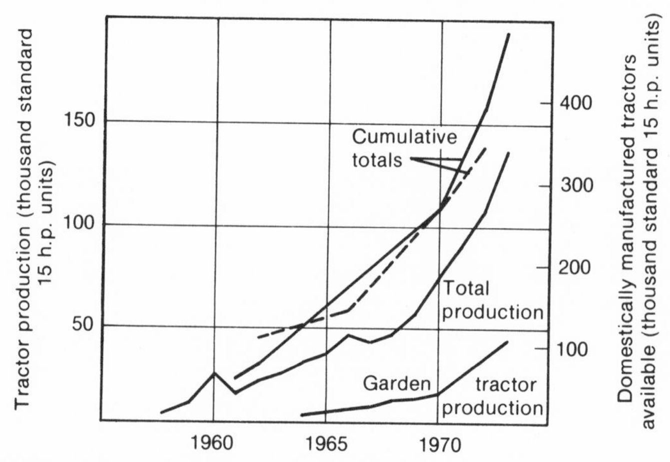

Sources:

Production of tractors (total and garden): Robert Michael Field, "Civilian Industrial Production in the People's Republic of China, 1949–74," in *China: A Reassessment of the Economy* (Washington, D.C.: U.S. Congress, Joint Economic Committee, 1975), p. 165. Field, a U.S. government analyst, does not indicate the basis for these estimates.

Cumulative totals: Computed (solid line) by adding up the domestically produced tractors in the previous five years, i.e., assuming that the tractors are fully depreciated after five years. This figure does not include imports, and therefore is artificially low for the period up to the middle or late 1960's. These two procedures make the estimated cumulative total quite conservative. Computations (barred line) from U.S. Central Intelligence Agency, *Production of Machinery and Equipment in the People's Republic of China* (Washington, D.C.: CIA, May 1975), p. 13.

Field's estimates, China was almost halfway there. One million standard tractors could be available by 1981 if present growth trends continue.

A few other fragments of evidence also suggest that the 1970's is the decade for mechanization. In 1974, the tractor-plowed area was nearly double that of 1965.[85] In 1964 it was reported that tractors cultivated about 10.7 million hectares.[86] If tractor-plowed area went up in 1965 by the same ratio that the number of tractors available went up (20 per cent; see Figure 9.1), then tractors would have plowed about 12.8 million hectares in 1965 and around 25 million hectares in 1974.

85. Chi Feng, "Thirteen Consecutive Years of Rich Harvests," *Peking Review* no. 1 (Jan. 3), 1975, p. 12.

86. Chao, *Agricultural Production*, p. 115.

In 1973, a secret document (captured by raiders from Taiwan) revealed that "the degree of agricultural mechanization . . . is now 27 percent. Efforts will be made to raise it to 40 percent by 1975." [87] Chinese economists had gauged that only about 80 to 100 million hectares could be mechanized (Table 7.8), so this would imply 22 to 27 million hectares are mechanized, virtually the same amount derived from the other data.

One other fragment of data confirms the plan to mechanize rapidly during the 1970's. In 1975 Chinese leaders projected the goal "to basically fulfill task of farm mechanization by 1980." [88] This plan is, presumably, part of the overall program outlined by Chou En-lai, "to build an independent and relatively comprehensive industrial and economic system" over the next ten years, "and to accomplish the comprehensive modernization of agriculture industry, national defense and science and technology before the end of the century." [89]

In a concluding report at the National Conference on Learning from Tachai in September–October 1975 (a conference designed to define rural policy for the late 1970's), Vice Premier Hua Kuo-feng concisely summarized China's policies toward agricultural mechanization:

> The development of farm mechanization will greatly raise labor productivity in agriculture and enable the peasants to set aside plenty of time to develop a diversified economy and build a new, prosperous and rich socialist countryside. It will also have a great significance in bringing into play the role of the people's commune as an organization that combines industry, agriculture, commerce, education and military affairs, in enabling the commune to display its superiority—big in size and with a high degree of public ownership—and in narrowing the differences between town and country, between worker and peasant and between manual and mental labor.
>
> Therefore, the various departments concerned under the State Council and the leading organs of the provinces, prefectures, and counties must make very great efforts to speed up the progress of this work so as to ensure that the great task of mechanizing agriculture will be accomplished in the main by 1980. [90]

87. Political Department, Kunming Military Region, *Reference Materials Concerning Education on Situation,* no. 45 (Kunming, April 6, 1973); available in *Issues and Studies* 10:10 (July 1974), 101.

88. "National Conference on Learning from Tachai in Agriculture," *Peking Review,* no. 38 (Sept. 19), 1975, p. 3.

89. Chou En-lai, "Report on the Work of the Government," to the First Session of the Fourth National People's Congress, Jan. 13, 1975; in *Peking Review,* no. 4 (Jan. 24), 1975, p. 24.

90. Hua Kuo-feng, "Mobilize the Whole Party, Make Greater Efforts to Develop Agriculture and Strive to Build Tachai-Type Counties Throughout the Country," *Peking Review,* no. 44 (Oct. 31), 1975, p. 10.

There is no reason to believe that Teng Hsiao-p'ing contradicted this approach. Yao Wen-yuan reportedly questioned Hua's mechanization policies. Yao challenged, "Can farm mechanization be realized in the main by 1980? I don't think [so]." Yao had reportedly blocked a *People's Daily* editorial on farm mechanization approved by Hua Kuo-feng in 1971, conceivably in conjunction with the second national conference on agricultural mechanization held that year. In December 1977, the third national agricultural mechanization conference was planned, and Mao's letter of March 12, 1966, which had inspired the first conference at Wuhan, was reprinted.[91]

To facilitate this tremendous expansion of the agricultural machinery industry, China systematized its tractor- and engine-production facilities. A new series of diesel engines was designed, based on the concept of interchangeable parts. Cylinder bores are 90, 95, 100, and 105 mm., and engines can be fabricated with from one to more than six cylinders, which may be horizontal or vertical, and air- or water-cooled. These engines are used in a wide range of tractors. In the early 1970's at least fifteen different tractor models were being manufactured in over a dozen localities.[92]

It should be pointed out that this program of tractorization complements other programs in agrotechnical transformation. During the 1960's and early 1970's vast improvements were made in the control of water in China. Chinese sources claimed that 33 million hectares were guaranteed harvests in spite of drought or flood. This includes 17 million hectares considered "freed from waterlogging,"[93] All together, China is thought to have had 40 to 44 million hectares irrigated in the early 1970's.[94]

This improvement in the irrigation system was made possible through two complementary programs: (1) mass mobilization of labor during the

91. Chou Chin, "Mechanization: Fundamental Way Out for Agriculture," *Peking Review*, Feb. 25, 1977, p. 15; "National Conference on Farm Mechanization Planned," NCNA Peking, Dec. 8, 1977; Mao Tsetung, "A Letter on Farm Mechanization," *Peking Review*, no. 52 (Dec. 26), 1977, pp. 7–9.

92. Detailed information is available in *T'o-la-chi Kuo-tsao*, JPRS 63,091, pp. 14–36; and Chi I-chai, "A Study on the Production and Need of Agricultural Machinery on China Mainland," *Fei-ch'ing Yueh-pao* 10:3 (May 31, 1967), in JPRS 42,271, pp. 17–19.

93. "Sharp Rise of Farm Machinery," *Peking Review*, no. 6 (Feb. 7), 1975, p. 23; Feng Nien, "How China Solved Her Food Problem," *China Reconstructs*, Jan. 1975, p. 6.

94. Dwight Perkins, "Constraints Influencing China's Agricultural Performance," in *China: A Reassessment of the Economy* (Washington, D.C.: U.S. Congress Joint Economic Committee, 1975), p. 360.

decade of the 1960's on rural public works, particularly canals, dikes, drainage systems, and land leveling;[95] (2) simultaneous mechanization of large parts of the irrigation and drainage systems. From 1965 to 1975 the horsepower of mechanical irrigation-drainage equipment rose from 8.6 million to 40–50 million. By 1975 there were 1.7 million mechanical tube wells serving over 7.3 million hectares.[96]

At the same time a vast commitment was made to expand very rapidly China's chemical fertilizer industry. During the 1960's and early 1970's about 1,000 intermediate-sized chemical fertilizer factories (making about 5,000–10,000 tons ammonia annually, which was then converted to ammonium bicarbonate) were constructed to supplement the large-scale fertilizer factories that had been built in the late 1950's.[97] By 1972 the large and intermediate factories together produced roughly 3.7 million tons of crop nutrients.[98] (This is an astonishingly large amount; it is more than Japan produces, and roughly three times what India produces.[99] But because of China's vast area, it does not result in high application rates.)

Between 1972 and 1974 China contracted with the U.S. Kellogg Corporation and with Dutch, Danish, French, and Japanese firms to purchase and construct fifteen large, modern ammonia factories which would go into production in the middle and late 1970's. These factories are together capable of producing about 3.78 million tons[100] of nitrogen per year, thereby raising China's already large supply of chemical fertilizer by about 65 per cent. China also purchased thirteen urea factories (some of which will be the largest in the world) to convert the ammonia into a fairly concentrated, easily transported solid fertilizer.[101]

There have been some significant advances in seed technology also. China has developed advanced methods of tissue culture to make some new varieties, has strengthened the extension system,[102] has imported

95. Jim Nickum, "A Collective Approach to Water Resource Development: The Chinese Commune System, 1962–72" (Ph.D. diss., University of California at Berkeley, 1974).

96. For 1965: Chao, *Agricultural Production,* p. 141. For 1975: "China Transforms Its Land and Rivers," NCNA Peking, Feb. 18, 1976, and "Cultural Revolution Spurs High-Speed Development of China's National Economy," NCNA Peking, June 29, 1976.

97. Jon Sigurdson, "Rural Industrialization in China," in *China: A Reassessment,* pp. 417–418.

98. Stavis, *Making Green Revolution,* p. 43. 99. *Ibid.,* p. 70.

100. Sigurdson, "Rural Industrialization," p. 417.

101. Hans Heymann, Jr., "Acquisition and Diffusion of Technology in China," in *China: A Reassessment,* pp. 726–727.

102. Benedict Stavis, "Research and Extension in China," *World Development* (1978).

Figure 9.2. Regions of modernizing agriculture, 1974. Reprinted with revisions, from *Bullet[in] of Concerned Asian Scholars* 7:3 (July–Sept. 1975), p. 25.

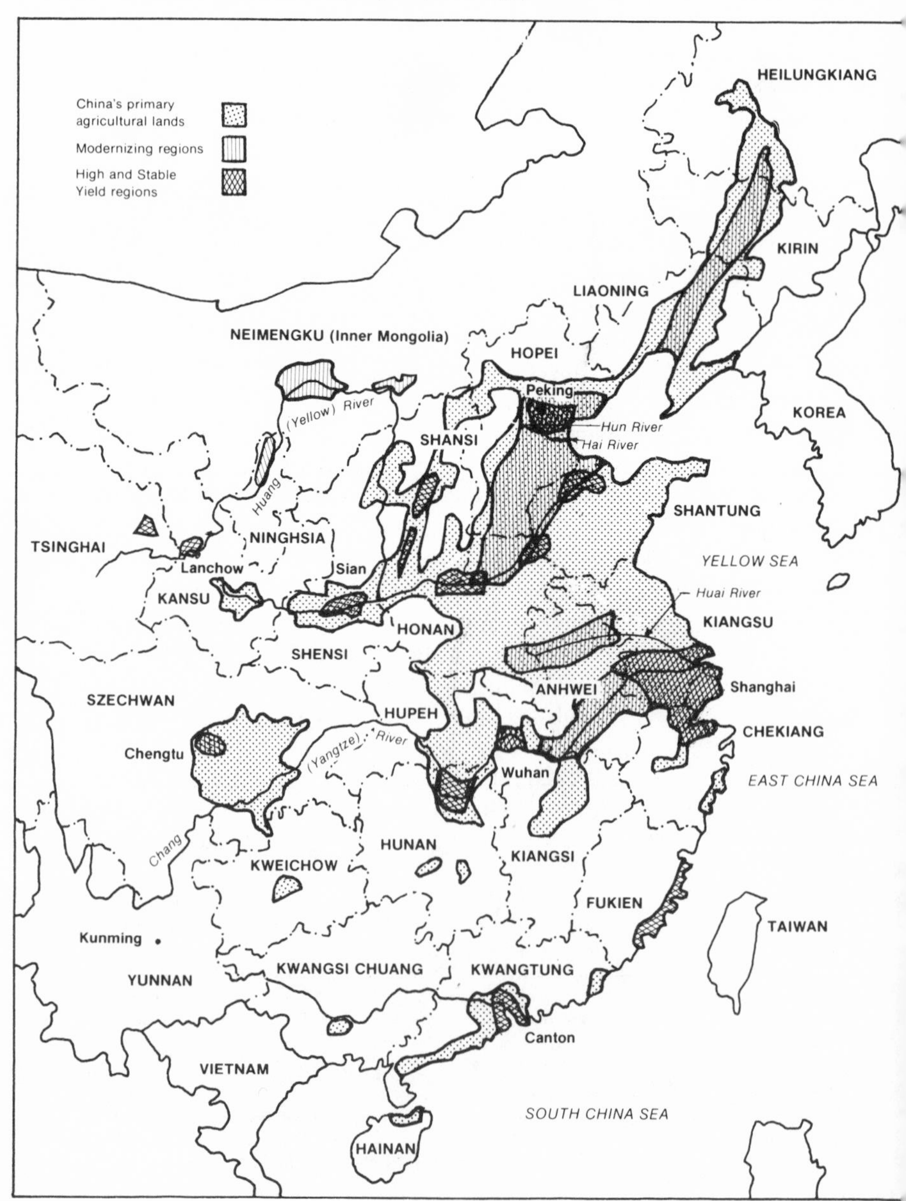

over 20,000 tons of high yielding Mexican varieties of wheat seed in 1973–75, and had resumed international communication on agricultural science and technology.[103] Visiting U.S. agricultural scientists, however, still had doubts about the adequacy of central research capabilities.[104]

One of the reasons that mechanization (and expansion of the chemical fertilizer industry) could proceed so rapidly was the phenomenal growth in China's petroleum industry. In 1957, Caho Hsüeh had estimated that tractorization would require roughly 4 million tons of diesel oil a year.[105] Not until 1960 did China extract over 4 million tons of oil products; and as late as 1968, 4 million tons constituted over one-fourth of domestic supplies. Petroleum extraction grew very rapidly in the late 1960's and 1970's through the opening of several oil fields, and by 1974, 4 million tons of oil represented only 6 per cent of supplies.[106] The tremendous expansion in the petroleum industry also provided the supplies for China's burgeoning chemical fertilizer industry.

An estimate of regions which have modernized agriculture significantly is made on Figure 9.2. Detailed sources for these estimates are available elsewhere.[107] While it is not possible to specify with precision the location of the roughly 25 million hectares thought to be machine-plowed in 1974, scattered data assembled for Table 9.4 indicate the general pattern. There is extensive mechanization in the Northeast. Major suburban areas are almost completely mechanized. On the North China Plain, mechanization is extensive. In South China, mechanization is proceeding more gradually but will undoubtedly accelerate with the expansion in production of small tractors.

The close relationship between mechanization and more intensive

103. *Plant Studies in the People's Republic of China* (Washington, D.C.: National Academy of Sciences, 1975), esp. pp. 56–57.

104. The full report of the visiting U.S. scientists is *ibid.* Summaries are available: Sterling Wortman, "Agriculture in China," *Scientific American* 232:6 (June 1975), 13–21; and G. F. Sprague, "Agriculture in China," *Science* 188:4188 (May 9, 1975), pp. 549–555.

105. This was based on the assumption that China would require 400,000 of the crawler-type tractors, and that each machine would consume over 10 tons of fuel a yea. Chao Hsüeh, "The Problem of Agricultural Mechanization in China," *Chi-hua Ching-chi,* no. 4 (April), 1957 (ECMM 87 p. 11).

106. Bobby Williams, "The Chinese Petroleum Industry; Growth and Prospects," in *China: A Reassessment,* p. 228. Another excellent source on this issue is Tatsu Kambara, "The Petroleum Industry in China," *China Quarterly,* no. 60 (Oct.–Dec. 1974), 699–719.

107. The map was published first in my article "How China Is Solving Its Food Problem," *Bulletin of Concerned Asian Scholars* 7:3 (July–Sept. 1975), 25. Data on which the map was compiled is available in "A Preliminary Model for China's Grain Production, 1974," *China Quarterly,* no. 65 (March 1976), 87.

Table 9.4. Regions and types of mechanized agriculture

Type and region	Amount mechanized	Date
Tractorization		
Northeast	one-third	1965 [a]
Heilungkiang	50%	1975 [b]
Peking suburbs	60%	1966,[c] 1974 [d]
Tangshan Special District (Hopei)	40%	1972 [e]
Hopei (whole province)	30%	1961 [f]
Shantung (whole province)	25%	1965 [g]
Honan	40%	1971[h]
Shanghai suburbs	88%	1975[i]
Sinkiang		
State Farms	80%	1975[j]
Communes	50–60%	1975[j]
Mechanical rice transplanting, seed drilling		
Hunan, rice transplanting	around 15,000 ha. (i.e. experimental, in a few localities)	1970[k]
Kwangsi (Luchuan County), rice transplanting	50%	1970[l]
Shanghai suburbs, late rice crop	44.5%	1975[i]
Peking, wheat drills	80%	1974[d]
	90%	1975[m]
Mechanized or semimechanized threshing		
Hunan	80% of rice	1973[n]
Pearl River Delta	Widespread	1972[o]

Sources:

a. "China Extends Range of Tractors, NCNA Peking, Jan. 31, 1966 (SCMP 3631, p. 19).

b. "Large-scale Agricultural Development Going on in Northeast China," NCNA Harbin, Jan. 19, 1976.

c. "Peking and Surrounding Country Districts Develop New Relationship," NCNA Peking, Feb. 7, 1966 (SCMP 3635, p. 22).

d. "Diversified Economy Thrives on Peking's Outskirts," NCNA Peking, Oct. 18, 1974.

e. My own observation, 1972.

f. NCNA Peking, May 13, 1961 (FBIS May 25, 1961).

g. Lowell Dittmer, "Shantung," in Edwin Winckler, ed., *A Provincial Handbook of China,* forthcoming.

h. Yu Wen, "How Honan Achieves Self-Sufficiency in Grain," *Peking Review,* no. 40 (Oct. 6), 1972, p. 27.

i. "Farm Mechanization on Shanghai's Outskirts," *Peking Review,* no. 40 (Oct. 3), 1975, p. 30.

j. Audrey Topping, "China Pushing Drive to Mechanize Farms," *New York Times,* Dec. 4, 1975.

cropping systems is apparent in a report from Shanghai. Between 1965 and 1975 a region which had been basically double-cropping was transformed to triple-cropping (on 90 per cent of the land sown to grain). In 1975 tractors plowed 88 per cent of the land. Mechanical rice transplanters planted 44.5 per cent of the late rice crop. Research and development was being done on small-scale harvesting machines.[108] Undoubtedly this mechanization was crucial in saving time to enable the third crop.[109]

By 1973, China was using somewhat more field machinery than most other Asian countries. Table 9.5 shows that the availability of horsepower per cultivated hectare in China was 50 per cent more than in India, 20 per cent more than in Pakistan, 75 per cent more than in Thailand. Even the level of mechanization planned for 1980, however, is still fairly low by the standards of Europe and other industrialized countries. By 1980, China's ratio of tractors to cultivated land will be only half of Italy's in the early 1960's, and far lower than North Korea's in 1973.

Management

After the Cultural Revolution commune ownership was obviously the preferred pattern of control for agricultural machinery, but institutional change was slow.[110] For a while there were no firm instructions on how to proceed on the question of tractor ownership. Some communes considered a May 7, 1966, directive (originally a letter from Mao to Lin Piao)

108. "Farm Mechanization on Shanghai's Outskirts," *Peking Review*, no. 40 (Oct. 3), 1975.

109. A detailed description of the types of agricultural machinery available in 1975, their contribution to agricultural production, and the factories that produced them is available in the report of twelve scholars who visited China in summer 1975 under the auspices of the U.S. National Academy of Sciences, specifically to study rural small-scale industry. The report is Dwight Perkins, ed., *Rural Small-Scale Industry in the People's Republic of China* (Berkeley: University of California Press, 1977). Chapter 5 deals with agricultural machinery.

110. The state farm system seems to have escaped criticism, perhaps because of its close relationship with the army and its importance in sensitive border regions.

---

k. Computed from Stavis, *Making Green Revolution*, p. 50.

l. Roland Berger, "Mechanisation of Chinese Agriculture," *Eastern Horizon* 11:3 (1972), 16.

m. Dwight Perkins, ed., *Rural Small-Scale Industry in the People's Republic* (Berkeley: University of California Press, 1977), p. 128.

n. "Machine-Building Industry," *Peking Review*, no. 44 (Nov. 2), 1973, p. 22.

o. My interviews, 1972.

Table 9.5. Field machinery in selected Asian countries, 1973, compared to figures for Italy (an industrialized country), 1961–65

Country	30-h.p. tractors (thousands)	5-h.p. tractors (thousands)	Total tractor h.p. (thousands)	Cultivated land (thousand ha.)	H.p./ha.
China (1973)[a]	200	253	7,120	127,000	.056
(1980)[b]	500	600?	18,000	135,000	.133
India[c] (1974)	200	10	6,050	160,610	.038
(1979)	500	100	15,500		.097
Pakistan	30		905	9,385	.047
South Korea	.01	4,900	29	2,241	.013
North Korea	23		690	1,894	.364
Thailand	13		398	12,431	.032
Bangladesh	2		66	8,900	.007
Sri Lanka	8		243	895	.271
Italy (1961–65)	78	62	2,667	9,511	.28

Sources: Unless otherwise specified, data are from United Nations Food and Agriculture Organization, *Production Yearbook, 1974* (Rome: FAO, 1975).

a. Computed from: U.S. Central Intelligence Agency, *Production of Machinery and Equipment in the People's Republic of China* (Washington, D.C.: CIA, May 1975), pp. 13–14. Data given originally in terms of 15-h.p. regular tractors and 4-h.p. (drawbar) garden tractors.

b. Projections based on plan basically to mechanize agriculture by 1980.

c. Government of India, Planning Commission, *Draft Fifth Five-Year Plan, 1974–79* (New Delhi: Government of India Press, 1974), II, 14.

sufficient indication of the new policy.[111] However, this judgment required considerable imagination because the directive merely stated in general terms that communes should, where appropriate, run small factories. Nothing was said specifically about tractors.[112] No communes referred to Mao's letter of March 12, 1966, which criticized the trusts; presumably this letter was not made public. Several references confirm that no directives were issued that dealt specifically with tractor ownership. Probably in late 1966, tractor station personnel at one location stated: "In the absence of directives from the higher level and without experience, we are afraid that in the event of trouble the responsibility would be too heavy for anybody. It would be better to wait for directives from the

111. Peikou Commune Tractor Station, P'englai County, Shantung, "Build the Commune Tractor Station into a Big School of Mao Tse-tung's Thought along the Road of the 'May 7' Directive," *Nung-yeh Chi-hsieh Chi-shu*, no. 9 (Sept.), 1968 (SCMM 632, p. 21).

112. "Letter to Comrade Lin Piao, May 7, 1966," *Chairman Mao on Revolution in Education* (People's Publishing House, 1967), CB 888, p. 17.

higher level.''[113] Even as late as 1968, tractor station personnel in some locations refused to assist the decentralization of tractors: ''They even obstinately cling to the line of state monopoly under the pretext that 'the discussion of the matter should wait until the struggle-criticism-transformation is over' and that 'no action is called for until the Central Committee has made known its attitude.' ''[114]

In middle and late 1968, articles appeared describing the experiences of decentralizing agricultural machinery at several communes and counties (see Table 9.6). Although these stories may have been representative of the general situation, it seems more likely that they were conscious experiments. It is clear that these reports defined the desirable pattern of ownership and management and served to instruct communes and brigades in the handling of specific problems that arise when they own and manage their own machinery.

The process of decentralization in these locations was cautious and careful, with great attention paid to various details of organization, personnel, and finance. It stands in sharp contrast to the very rapid decentralization program of 1958. Indeed, the former experience seemed to be recalled as a negative model. Personnel at one tractor station reminded everyone: ''When tractors were sent to the communes in 1958, certain loss was suffered because China's Khrushchev employed counterrevolutionary double-dealing tactics.''[115]

From these various reports the important elements of the new policy can be pieced together.

Widespread experimentation. The first feature of the decentralization after the Cultural Revolution was the relatively widespread nature of the experimentation. In the decentralization experience of 1957–58, preliminary experiments had been confined to four cooperatives in two locations: near Peking and in Heilungkiang. In contrast, in 1966–68, experiments were reported in at least ten provinces. (Remarkably, no experiments were reported in the Northeast; the reasons for this apparent gap are unknown.) Experiments were reported on both commune and county levels.[116]

113. Chengting County Agricultural Machine Work Station, Hopei, ''March with Big Strides on the Revolutionary Road—How Chengting Hsien Station's Farm Machines and Tools Were Sent Down,'' *Nung-yeh Chi-hsieh Chi-shu,* no. 11 (Nov. 8), 1968 (SCMM 643, p. 8).

114. ''Everything Based on the Interests and Needs of the Poor and Lower-Middle Peasants,'' *Nung-yeh Chi-hsieh Chi-shu,* no. 6 (June 8), 1968. (SCMM 620, p. 27).

115. ''March with Big Strides,'' SCMM 643, p. 8.

116. *Ibid.*

Table 9.6. Reports of experiments and models for decentralization of tractor ownership, 1965–68

Province	Commune, county	Date of experiment	Inspiration for experiment
Kwangtung [a]	Lengk'eng Commune, Huaichi County	1965–66	Poor peasants, commune party
Inner Mongolia [b]	Linho County	1967	Communes, brigades, teams; Cultural Revolution critiques
Kansu [c]	Yungten County	1966–68	County, commune
Hupeh [d]	Hsinchou County	1965–66	Poor peasants, masses
	Hsukuang Commune, O-ch'eng County	Before 1968	Brigade
Kiangsu [e]	P'aoche Commune, Hsinyi County	1965–66	Commune
Shantung [f]	Peikuo Commune, P'englai County	1966–67	Poor peasants, May 7 directive
Shansi [g]	Jui-ch'eng County	April–Nov. 1968	County revolutionary committee
Hopei [h]	Chengting County	Dec. 1966–May 1968	MTS employees, poor peasants
Honan [i]	Lank'ou County	Feb. 1968	County revolutionary committee
Shansi [j]	Chaochia hsiao-ts'un Commune	Jan.–March 1968	Poor peasants, May 7 directive

Sources:

a. Agricultural Mechanization Office of Lengk'eng Commune, Huaichi County, Kwangtung, "Initial Experience of Lengk'eng Commune in Promoting Agricultural Mechanization on a Large Scale," *Nung-yeh Chi-hsieh Chi-shu,* no. 4 (July 8), 1967 (SCMM 600, pp. 8–16).

b. Wang Shih-ch'un, Linho County Revolutionary Committee, Inner Mongolia, "Nothing Can Stand in the Way of Mass Enthusiasm for Farm Mechanization," *ibid.,* no. 2, 1968 (SCMM 620, pp. 12–13).

c. Yang Heng, P'u Pa-shan, Chao Tzu-yün, and Pai Yung-fa, "The People's Commune Can Achieve Farm Mechanization with Greater, Faster, Better and More Economic Results—Refuting the Theory that Mechanization Must Go through the Stage of State Operation," *ibid.,* no. 8 (Aug. 8), 1968 (SCMM 629, p. 39).

d. Revolutionary Leading Team, Hsinchou Farming Machinery Bureau, Hupeh, and Hsinchou New Farming Machinery Revolutionary Rebel Field Corps, "Farm Mechanization Promoted by Communes in Hupeh's Hsinchou County along the Revolutionary Route Charted by Chairman Mao," *ibid.,* no. 9 (Sept.), 1968, supplement (SCMM 630, pp. 29–36): Production Brigade No. 2, Hsukuang Commune, O-ch'eng County, Hupeh, "The Question of Mechanization by the Collective," *ibid.,* no. 9 (Sept.), 1968, supplement (SCMM 644, pp. 13–19).

e. Production Committee of P'aoche Commune, Hsinyi County, Kiangsu, "Farm Mechanization through Communes," *ibid.,* no. 2, 1968 (SCMM 620, pp. 14–20).

Local planning. The experiments in 1966–68 appear to have been more autonomous of central government leadership than those in 1957–58. The earlier experiments had been carried out under the direct supervision of K'ang Sheng after he returned from a tour of the Soviet Union and Eastern Europe. After one year, the results of the experiments were reported at a major conference of the agricultural machinery bureaucracy, and instructions were soon issued ordering decentralization.[117] In 1966–68, the experiments were apparently conducted on a less formal basis. There is no indication that they were directed by the central government, and it is interesting to note that most were conducted *before* the establishment of a provincial revolutionary committee in the relevant province. The experiments were not, however, completely spontaneous. Many were inspired by the rather vague May 7, 1966, directive, by the Red Guard critiques in technical journals, and by the reissue of Mao's instructions from 1958 and 1959 regarding agricultural mechanization. Extensive reports published after the experiments were completed provided some concrete suggestions but permitted details of the process to be left to the various communes and counties. Of course, the decentralized character of these experiments may have been out of necessity—not choice—because at that time central authority had eroded to a considerable degree. It had not eroded completely, however, and it remains possible that experiments could have been directed by the central ministry.

County leadership. The experiments were generally conducted under

117. See Chapter 4 above.

---

f. Peikou Commune Tractor Station, P'englai County, Shantung, "Build the Commune Tractor Station into a Big School of Mao Tse-tung's Thought along the Road of the 'May 7' Directive," *ibid.*, no. 9 (Sept.), 1968, supplement (SCMM 632, pp. 21–24).

g. Revolutionary Committee for Jui-ch'eng County, Shansi, "Use Mao Tse-tung's Thought to Direct the Work of Sending down the Farm Machines and Tools of State Stations," *ibid.*, no. 11 (Nov. 8), 1968 (SCMM 643, pp. 3–7).

h. Chengting County Agricultural Machine Work Station, Hopei, "March with Big Strides on the Revolutionary Road—How Chengting Hsien Station's Farm Machines and Tools Were Sent Down," *ibid.*, (SCMM 643, pp. 8–12).

i. "It is Good for Tractors to Revert to Chairman Mao's Revolutionary Line—Report on Investigation of Change in Management of Tractors by the Collective in Lank'ou County, Honan," *ibid.*, no. 10 (Oct. 8), 1968 (SCMM 643, pp. 17–26).

j. Farm Machine Station of Chaochia Hsiao-ts'un Commune, Ta-t'ung City, Shansi, "Rely on the Masses to Run the Farm Machine Station Democratically," *ibid.*, no. 9 (Sept.), 1968, supplement (SCMM 644, pp. 7–12).

the leadership of the county revolutionary committee. In one county, the group that actually administered the experiment was headed by a committee vice chairman and composed of nine responsible persons of the agricultural, finance, trade, planning, and personnel departments of county organizations and the former state agricultural machine stations.[118] Another experiment was carried out "under the powerful leadership of the county revolutionary committee."[119]

After decentralization was completed, the county revolutionary committee retained important leadership functions. In one county it was reported:

> The former county agricultural machine general office was reorganized into an agricultural machine management station—a three-in-one administrative structure in charge of management, repair and supply—to be staffed with eight cadres and technical personnel. The station was responsible for all-around management of the mechanization in the county area and for scientific research in agricultural machinery. Under the station were set up a repair plant and an agricultural machine supply company.[120]

This continued leadership and assistance from the county level was in contrast to the situation in 1958 when, according to Red Guard groups, communes were given no technical assistance in managing and maintaining their agricultural machinery.

Dispatching machines. A major problem in "sending down" agricultural machinery was to dispatch the appropriate equipment to the proper commune or brigade. Both technical and political factors were involved. Various reports of experiments claimed that this problem was studied carefully, and rational choices were made:

> As to the localities to which [machines] were sent, the original working areas were taken as the base areas with consideration given to the important communes and teams. Aid was given to mountain areas with consideration given to poverty-stricken communes and teams. In addition, rational distribution of agricultural machines and implements was made with consideration given to the terrains of the communes and teams, repair force and the technical state of machines and implements.[121]

118. Revolutionary Committee for Jui-ch'eng County, Shansi, "Use Mao Tse-tung's Thought to Direct the Work of Sending down the Farm Machines and Tools of State Stations," *Nung-yeh Chi-hsieh Chi-shu*, no. 11 (Nov. 8), 1968 (SCMM 643, p. 4).

119. "It is Good for Tractors to Revert to Chairman' Mao's Revolutionary Line—Report on Investigation of Change in Management of Tractors by the Collective in Lank'ou County, Honan," *Nung-yeh Chi-hsieh Chi-shu*, no. 10 (Oct. 8), 1968 (SCMM 643, p. 17).

120. "Use Mao Tse-tung's Thought, SCMM 643, p. 5. 121. *Ibid.*, p. 6.

Thus it appears that the decentralization of 1966–70 was physical as well as administrative and financial, providing another contrast to the experience in 1958.

Personnel. In the 1966–68 decentralization experiments, considerable attention was devoted to the transferring of personnel, to assure both that decentralization was achieved and that the enthusiasm of experienced personnel was maintained. The general policy was that tractor station cadres, originally employed by the state, would be transferred to the commune payroll with the exception of those few retained at the county service unit. This involved changing from monthly wages to work points. In one location, however, there was a two-year "grace period," during which the personnel would work at the commune level but would continue to receive salaries and food rations from the state.[122] Another commune evolved a sophisticated combination in which a machine operator received two wages: "He receives his pay for his mechanical work from the mechanical team according to the average work point value of the brigade, and he receives his share of distribution from the production team where he does farm labor."[123] Work points from the mechanical team were to be based on a modified Tachai system, which included these criteria:

1. Puts politics to the fore, studies and applies Mao's works;
2. Does good job of propagating thought of Mao;
3. Observes rules and regulations, engages in production safely;
4. Loves productive labor ardently, practices industry and thrift, gives good services;
5. Unites with and helps others, in touch with masses, good attitude.

The plan called for quarterly assessment, with publication of the results.[124]

It is not clear how the pension rights and medical benefits of state workers were handled in the transfer, although they were discussed.[125] In one brigade, "if the mechanical personnel sustain injuries while on duty, the medical expenses will be borne by the collective, which will also pay the wages in the period of their confinement."[126] The brigade also

122. "It is Good for Tractors," SCMM 643, p. 25.

123. Production Brigade No. 2, Hsukuang Commune, O-ch'eng County, Hupeh, "The Question of Mechanization by the Collective," *Nung-yeh Chi-hsieh Chi-shu,* no. 9 (Sept.), 1968, supplement (SCMM 644, p. 15).

124. *Ibid.*

125. "Use Mao Tse-tung's Thought," SCMM 643, p. 6.

126. "Question of Mechanization by the Collective," SCMM 644, p. 19.

agreed to provide, at collective expense, protective clothing, soap, and towels.

This brigade made efforts to assure development of expertise and continuity of experience. It decided that "all personnel were fixed properly [to an assignment] at one time and would not be changed for the year." Other regulations indicated that labor discipline would be maintained under decentralized management:

> Do not allow non-mechanical personnel to drive the machines; do not allow people to work with bare feet; do not allow people to leave the machines after these have been set in motion; do not allow people who are on duty to doze off; do not allow people operating the machines to talk and laugh; do not allow people to work with electric wires trailing behind them; do not allow people to desist from wearing safety devices when working; do not allow people to shut the gate while carrying a load; do not allow people to dismantle the machine at will; and do not allow people to operate the machine if they are sick.[127]

Finance. When tractors were "sent down" in 1966–68, care was taken to ensure that financial and accounting arrangements were understood and considered fair by all parties concerned. In one county, the communes were charged for each piece of machinery at the current price (thereby giving communes the advantage of decreasing costs of manufactured goods), less depreciation. Depreciation was computed on the basis of the average between the current and original prices. Communes had two years to pay for the machinery. They were expected to raise the funds either from the economic accumulation fund or by apportioning charges to the teams on the basis of cultivated area. There was no mention of credit from any of China's banks. (However, in one commune, the wealthier brigades lent money to the poorer brigades.[128]) The funds were paid to the county agricultural machinery bureau, which was instructed, pending directives from higher levels, to retain much of it for establishing a service network. A special team was organized to assure that the procedures were followed carefully, that contracts were signed, and that the various teams, financial departments, banks, and agricultural machinery stations were carrying out the contracts.[129]

In another county the computation of the value of the machinery was described in detail:

127. *Ibid.*, pp. 14–19.

128. Farm Machine Station of Chaochia hsiao-ts'un Commune, Ta-t'ung City, Shansi, "Rely on the Masses to Run the Farm Machine Station Democratically," *Nung-yeh Chi-hsieh Chi-shu,* no. 9 (Sept.), 1968, supplement (SCMM 644, p. 9).

129. "Use Mao Tse-tung's Thought," SCMM 643, p. 7.

First, the machines and implements were technically appraised in detail. An appraisal and evaluation team was then formed to fix the price of each engine in light of the present prices of new engines. Finally a symposium of leading aged workers, cadres and technical personnel was held to examine the engines one by one and make appropriate corrections before submitting the prices to the hsien revolutionary committee for consideration.[130]

One commune specifically stated that the machines had to be fully repaired before being sent down.[131]

Commune preparation and mass participation. An essential element of the 1966–68 experiments was extensive preparation of the commune so that it would be ready to accept and manage the machinery. In one case, "a management committee for the farm machinery station was formed; this was a three-way-combination administrative organ composed of brigade cadres, representatives of poor and lower-middle peasants, personnel of the farm machinery station and revolutionary leading cadres at the commune level."[132]

Steps were taken to assure that these machine management committees were in close contact with both cadres and masses. Special study classes were conducted for commune cadres and personnel connected with agricultural machinery. Mao's directives on mechanization were studied, as were the experiences of neighboring communes.[133] Mass education and propaganda were essential too:

It was necessary to let the masses know why it was essential to set up the stations. To this end, we printed Chairman Mao's directives on agricultural mechanization and organized forces to publicize, through the media of meetings, black-board bulletins, broadcast stations, propaganda pictures and small programs, the momentous significance of commune-run agricultural machine stations, instances of mechanization found in the past years.[134]

Widespread discussions throughout the commune were expected to assist in many decisions:

As for all major problems confronting the farm machinery station, the management committee was to discuss them and submit preliminary plans. After that, the various committee members would report separately to their own brigades and solicit views and opinions from all quarters concerned. It was then that the management committee would make its decisions on the basis of the opinions of the masses and put them into effect with the approval of the commune. In this way,

130. "March with Big Strides," SCMM 643, p. 10.
131. "Use Mao Tse-tung's Thought," SCMM 643, p. 6.
132. "Rely on the Masses," SCMM 644, p. 7.
133. "March with Big Strides," SCMM 643, p. 11. 134. *Ibid.*

not only was it insured that the farm machinery station would be placed under the direct leadership of the commune, but democracy would be more fully practiced and the basic-level revolutionary cadres as well as the broad masses of poor and lower-middle peasants encouraged further to take a direct part in management. As a result, many problems were solved more smoothly.[135]

One example given was how mass participation facilitated rearrangement of fields to utilize better the machinery.

In order to raise the efficiency of tractors, to plow as many fields as possible, and reduce the travelling time of tractors, we again sought a solution to this problem by means of democratic discussions by the masses. . . . Very quickly the interlaced fields of various brigades were readjusted. The fields of various production teams were also re-arranged to suit the cultivation of crops. At the time of machine-farming, unified responsibility was taken by the brigade without being bound by restrictions caused by the boundaries between brigades and between fields. In this way, the efficiency of mechanical operations was considerably boosted.[136]

Democratic discussions were also useful in raising money to purchase the machinery. In one commune, after discussions among the various brigades, it was agreed that each should pay ¥1.4 for each mou of arable land. Wealthier brigades would lend impoverished brigades money to make this investment, and these loans would be repaid from the income of the commune's tractor station. (No mention was made of interest payments in connection with these loans.)[137]

While cadre and mass education were both important, they did not fill the need for specialized personnel at the commune level to manage the machinery. In one commune, "under the [revolutionary] committee was a commune agricultural machinery station which was responsible for commune management. The agricultural machinery station had one station master, one technical deputy master, one bookkeeper and one storekeeper, all of whom were partially detached or not detached from production pursuits."[138]

Although the Chinese government has not released statistics on tractor ownership in the period after the Cultural Revolution, an official concerned with agricultural mechanization mentioned informally in the spring of 1972 that the transition to the commune ownership system had been "basically" completed.[139] Moreover, interviews with personnel at

135. "Rely on the Masses," SCMM 644, pp. 7–8. 136. *Ibid.*, p. 10.
137. *Ibid.* 138. "Use Mao Tse-tung's Thought," SCMM 643, p. 5.
139. Benedict Stavis, Diary, April 28, 1972.

commune-run tractor stations near Peking confirm that the policies suggested in the Chinese journals were generally followed. Specialists who visited China in 1975 reported that the machinery was well maintained, spare parts were avilable, and utilization rates were high—roughly 2,000-2,500 hours per year, which is 4 times as high as in Japan and 2.5 times as high as in the United States.[140] Apparently commune management has avoided the inefficiencies which occurred in 1958.

Although the apparently smooth decentralization of tractors after the Cultural Revolution was primarily due to careful administration and a favorable political climate, one other factor should be mentioned. In 1958, the Soviet Union disbanded its machine tractor stations, selling machinery to the kolkhozy. Because the Chinese tractor system had closely paralleled the Soviet one, the Chinese personnel must have been much better prepared psychologically in the late 1960's for decentralization. It was, after all, continuing the Soviet model, and various procedures had been worked out there for handling the administrative problems of decentralization.

It should be pointed out that although the trusts were dropped in *management* of agricultural machinery, they were not totally abandoned in the *production* aspect. For example, in Tangshan a porcelain corporation integrated the activities of a variety of factories involved in porcelain; the Kirin Chemical Company did the same for chemical factories in the city of Kirin.[141] In agricultural mechanization, although the complex network of central, province, county, and commune factories does not appear to be part of an integrated trust,[142] the manufacture of a few special parts, such as engines, might be under quite strict central government coordination, in a fashion similar to the trusts. China has remained sensitive to the advantages of centralized organization and is selective about where such a pattern should be applied.

While in most dimensions of agricultural mechanization (such as soil preparation, irrigation, processing) the commune management system seems efficient, mention should be made of one type of problem which may eventually arise. For mechanized harvesting of grain, the commune ownership system may turn out to be an impediment to technical transfor-

140. Perkins, *Rural Small-Scale Industry,* p. 151.
141. "Pollery and Porcelain of Tangshan," *Peking Review,* no. 32 (April 11), 1972 p. 22; "Kirin's Chemical Industry Makes Some Changes," *China Reconstructs,* Aug. 1972, p. 20.
142. Perkins, *Rural Small-Scale Industry,* p. 142–149.

mation. The main problem with mechanized harvesting, of course, is that application of machinery is very difficult where complex intercropping techniques are used. Another issue involves management. Unless China develops much smaller-scale combine harvesters than are now available (and China is, indeed, doing research on this line), the economies of scale in combine harvester operation require operation on farm lands larger than those of a single commune. If operated on a single commune, the utilization rates remain very low. At Red Star Commune, for example, combine harvesters are used only five days a year.[143] A more efficient mode of using combine harvesters would be to have them move north with the harvest. One could imagine communes contracting harvesting services to neighboring (particularly northern and southern) communes. An even more efficient utilization of machinery might be arranged by a province owning a fleet of combine harvesters which moves north with the harvest and charges a fee for service. Such a pattern of management seems close to the trust system, however, and political problems would be raised by its adoption.

Personnel

During the immediate post–Cultural Revolution period the personnel appointments and administrative changes were consistent with a policy of emphasis on agricultural mechanization. In 1970, Sha Feng was appointed Minister of Agriculture and Forestry. Sha had a military background, and was a tank expert. While his background in agriculture may have been weak, he certainly was knowledgeable in setting up a system to make efficient use of internal combustion machinery. Moreover, he might have had close ties to agrotechnical personnel, because since the early 1950's demobilized tank drivers have become tractor drivers, especially in the West and Northeast state farms. Some Pekingologists interpreted his appointment as an indication of Lin Piao's military associates taking over the civilian administration from Chou En-lai. However, in view of his previous technical experience, this interpretation seems incomplete. Sha Feng survived Lin Piao's demise politically but has not been prominent.[144]

Other high-level appointments after the Cultural Revolution also indi-

143. Sid Engst, public address at Cornell University, Oct. 10, 1975.

144. A thoughtful survey of military-civilian relations is Ellis Joffe, "The PLA in Internal Politics," *Problems of Communism* 24:6 (Nov.–Dec. 1975), 1–12.

cated that agricultural mechanization was not forgotten. One of the vice ministers of Agriculture, Yang Li-kung, had been deputy director of the Loyang tractor factory and then a vice minister in the Ministry of Agricultural Machinery and the Eighth Ministry of Machine Building. His presence at a high level in the Ministry of Agriculture assured that technical expertise and historical continuity would not be lost as a result of the Cultural Revolution.

After the Cultural Revolution, the Eighth Ministry of Machine Building was merged with the First Ministry of Machine Building.[145] One of the "responsible personnel" in the new First Ministry of Machine Building, Hsü Pin-chou, had previously been a vice minister of the Ministry of Agricultural Machinery and the Eighth Ministry of Machine Building and had written an article stressing the importance of semimechanization in 1964.[146] Thus there was an element of continuity in the machine-building ministries. The Second Ministry of Light Industry, which had been set up in 1965 to organize a semimechanization campaign, was probably merged with the Ministry of Light Industry; some personnel (e.g., Hsieh Hsin-ho) continued to work in this sector, providing continuity.[147]

Rural Industry

Agricultural mechanization policy has complemented the emphasis on rural local industries.[148] The statements at the Chengtu Conference in 1958 stressed the relationship. Agricultural mechanization would provide the stimulus and market for rural industry; rural industry would provide the equipment for farm mechanization. China is not the only place to have discovered the critical role farm mechanization can play in develop-

145. Tillman Durdin, "China Simplifies Ministry Line-up," *New York Times,* Oct. 26, 1970, p. 6.

146. Hsü Pin-chou, "Struggle for the Realization of Agricultural Mechanization in China," *Kung-jen Jih-pao,* Oct. 10, 1964 (SCMP 3329, p. 9). Biographic material on Hsü is from the files of Professor Donald Klein. Hsü was identified as a responsible person of the First Ministry of Machine Building at a symposium on rice-transplanting machines in July 1970; this seems to indicate that he was still concerned with agricultural mechanization. Changsha, Hunan, radio, July 24, 1970.

147. Biographic data from the files of Donald Klein.

148. Excellent studies on rural industrialization in China include Perkins, *Rural Small-Scale Industry,* and several articles by Jon Sigurdson: "Rural Industrialization," in *China: A Reassessment;* "Rural Industry and the Internal Transfer of Technology," in Schram, ed., *Authority, Participation and Cultural Change in China,* pp. 199–232; "Rural Industry—A Traveller's View," *China Quarterly,* no. 50 (April–June 1972), 315–337; "Rural Industrialization in China: Approaches and Results," *World Development* 3 (1975), 527–538.

ing rural industry. In the Pakistan Punjab, Tanzania, and other places the same discovery is being made.[149]

Sound economic logic underlies the particular role farm mechanization can play in stimulating rural local industry.[150] The manufacturing processes of farm machinery (with the exception of a few parts) seem to involve few economies of scale. In Pakistan, large-scale engine and pump factories are simply multiples of small-scale factories. The technology and economics of production do not vary with plant size.[151] Moreover, consumers of farm machinery are not overly concerned with the finish of the product, a characteristic which is more easily attained in large, capital-intensive factories; farmers are more concerned that the machine has been designed to meet local requirements, and that it can be easily maintained and repaired. Another factor is transportation costs. Farm machinery is rather bulky and heavy; transportation costs can be substantial if the machinery is manufactured at some distance from the users. Rural local factories can excel in all these regards.[152]

Another benefit of local manufacture of farm tools and machinery is that this is an ideal stimulant for local industry. Most parts of farm tools do not need high grades of steel or very precise machining. Local deposits of ore can be used, and local refining processes developed. Unskilled labor can learn on machine tools which are not precise enough for other types of manufacture. Eventually, these local skills, materials, and

149. Frank Child and Hiromitsu Kaneda, "Links to the Green Revolution: A Study of Small-Scale, Agriculturally Related Industry in the Pakistan Punjab," *Economic Development and Cultural Change* 23:2 (Jan. 1975), 249–275; David Vail, *Technology for Ujama Village Development in Tanzania,* Syracuse University, Foreign and Comparative Studies, Eastern Africa Series 18 (1975).

150. Perkins (*Rural Small Scale Industry,* pp. 56–116) refrains from an overall judgment on the economic efficiency of small-scale plants, but gives much data on their productivity and indirect benefits.

151. Child and Kaneda, "Links," p. 269.

152. Two excellent analyses of the economic dimensions of rural, labor-intensive industry are David Vail, "The Case for Rural Industry: Economic Factors Favoring Small Scale, Decentralized, Labor-Intensive Manufacturing" (unpublished paper, July 1975); and David Morawetz, "Employment Implications of Industrialization in Developing Countries: A Survey," *Economic Journal* 84: (1974), 491–542.

A careful analysis of low-cost technology for tube wells in East Pakistan shows that it had a lower need for imported components, had higher rates of return on investment, provided important training for foremen and drillers, generated a major stimulus for small-scale industry, and provided a better-managed irrigation system because farmers understood and could maintain the technology. John Woodward Thomas, "The Choice of Technology for Irrigation Tubewells in East Pakistan: An Analysis of a Development Policy Decision," in Peter Timmer et al., eds., *The Choice of Technology in Developing Countries* (*Some Cautionary Tales*) (Cambridge: Harvard Center for International Affairs, 1975), pp. 41–50.

equipment can be upgraded and diversified from farm tools to other products.

In China, the general program calls for the development of five types of rural industry; (1) power, including hydroelectricity and local coal mining, to provide energy for local industry; (2) iron and steel, which will provide materials for local machine-building industries; (3) chemical fertilizer and other agricultural chemicals, to improve yields; (4) cement, for use in construction projects; and (5) farm machinery repair and manufacture.

For the latter, an integrated network of factories is being established. Small repair shops, often at brigade level, may have a lathe, a drill press, and perhaps welding equipment; they may be able to manufacture moderately sophisticated semimechanized equipment such as animal-drawn seed drills and threshers. Commune workshops are much more sophisticated and can make most repairs and major overhauls. Commune repair shops which I visited in 1972, presumably some of the more advanced, were able to tear down tractor engines, replace rings, bearings, valves, etc., work on diesel injectors and pumps, solder radiators, and weld parts. Some commune repair shops begin to manufacture farm tools, such as rice transplanters, horse-drawn harvesters, etc. Commune factories with forges and foundries can manufacture a small tractor, provided the engine and some other specialized parts are supplied.

County factories are supposed to be able to carry out any repair on any machine in the county. They often can make tractors, electric motors and pumps, food-processing machines, fodder choppers, etc. County factories also can make lathes, ball bearings, and other crucial components in the machine-building industry. One careful analyst estimates that there may be about 3,000 county-run machinery plants in China.[153]

By 1977 there were over a million brigade and commune factories, employing 17 million people,[154] about 6 per cent of the rural labor force.[155] This figure is consistent with scattered reports summarized in Table 9.7.

These local factories are absolutely vital in producing farm machinery.

153. Sigurdson, "Rural Industrialization," in *China: A Reassessment,* p. 412.

154. "Commune- or Brigade-Run Enterprises," *Peking Review,* no. 4 (Jan. 27), 1978, p. 30. Sigurdson's estimates for the early 1970's were 500,000 factories with 10 to 17 million workers. "Rural Industrialization," p. 412.

155. Sigurdson estimates the total labor force at 350 million by taking 70 per cent of the population aged 15–64. Presumably about 80 per cent of these people live in the countryside, so the rural labor force is about 280 million.

Table 9.7. Reports on rural industry

Location	Rural industry employment		Value rural industry product			Value added by agricultural side-line workers	Extent of mechanization
	No.	%	Yuan	% of local production	Production /worker		
Huatung Commune, Kwangtung [a]	828	4	¥ 553,400	10	¥ 670	¥294	223 tractors, rice threshing
Commune near Tsinan, Shantung [b]		20		55			80% tractorized, threshing, fodder crushing
Lin County, Honan [c]		3	¥21,000,000		¥2,630		
Kung County, Honan [d]				32			
Tsunhua County [c]	13,500	7	¥18,000,000	20	¥1,330	¥360	
Commune near Wuhan [e]	800	14	¥ 1,000,000+	40	¥1,250+	¥446	Extensive

Sources:

a. Ward Morehouse, "Notes on Hua-tung Commune," *China Quarterly* no. 67 (Sept. 1976), pp. 590–591.

b. Lu Hsuan and Chou Chin, "People's Communes, Establishment and Development," *Peking Review,* no. 34 (Aug. 23), 1974, p. 17.

c. Jon Sigurdson, "Rural Industry—a Traveller's View," *China Quarterly,* no. 50 (April–June 1972), pp. 315–337.

d. "Chinese Peasants Run Industries," NCNA Chengchow, Sept. 26, 1975.

e. "People's Communes Provide Jobs for All Able-bodied Peasants," NCNA Wuhan, Sept. 24, 1975.

In 1970, 80 per cent of the total value of farm machinery made in China was composed of medium-sized and small machines produced locally.[156]

Rural local industry is profitable. Although using 6 per cent of the labor force, rural industry generated 23.1 per cent of commune, brigade, and team income.[157] Table 9.7 shows similar ratios of higher labor productivity in rural industry. This, of course, is to be expected, as industrial workers have more capital and machinery at their disposal. Wages for rural industrial workers are not, however, three to four times higher than wages for farm labor. Rather, they are only slightly higher, roughly ¥40 per month instead of about ¥30 per month for agricultural workers (from collective sources). The profits generated by rural local industry can be used to provide capital for further expansion or to subsidize welfare services (especially health and education for the local community). In 1976, commune industry in fourteen provinces amassed ¥3.6 billion, of which 36 per cent was used to finance agriculture. In Hopei, commune industry provided eleven times more funds for agriculture than did state investment.[158] Much of the responsibility for entrepreneurship rests in local governments, whose various levels (brigades, communes, counties) identify needs, consider markets, assess the availability of resources, raise investment funds, and carry out the development of local industry. (Had the trust system been adopted, more entrepreneurial responsibility would have been in the central trusts, and local governments would have had less responsibility.)

The Chinese do not believe that rural local industry is the only element in industrialization. The general policy is described by the slogan "walk on two legs," of which rural local industry is one leg. A system of modern, capital-intensive, urban factories complements the rural factories. These manufacture the complex precision parts (engines and parts of engines), and supply tool steel and special alloys for the rural factories. In addition, they may pass down old machinery and train the local industry work force. The Chinese leadership believes that "walking on two legs"—developing complementary relationships between capital-intensive and labor-intensive industries—requires socialist planning. This includes a strong role for the state to fix prices, control investments, control

156. Roland Berger, "The Mechanisation of Chinese Agriculture," *Eastern Horizon* 11:3 (1972), 12.

157. "Commune- or Brigade-Run Enterprises," Peking Review, no. 4 (Jan. 27), 1978, p. 30. Also, Perkins, *Rural Small-Scale Industry,* pp. 40–41.

158. "How China Raises Funds for Agricultural Development," NCNA Peking, Nov. 30, 1977.

foreign economic penetration, downgrade profit incentives, and encourage exchange between factories. In the absence of these policies, the Chinese fear that the small-scale rural industries might not withstand the competition from modern industry. However, complementary relationships between factories of different scales and technologies have been noted in places without socialist economic planning (including Hong Kong, Taiwan Province, Mexico, Singapore, South Korea, and regions of India and Pakistan),[159] suggesting that a variety of political-economic climates permit the development of small-scale rural industry.

Lin Piao and Agricultural Mechanization

For a brief period after the Cultural Revolution agricultural mechanization policy apparently became intertwined with political conflict at the highest level. Scattered data suggest that Lin Piao opposed the general policy of taking agriculture as the base of the national economy, of orienting the industrial system to supply agricultural inputs, and of supplying equipment for agricultural mechanization.

Evidence of continued high-level debates came from Ch'en Yung-kuei, the leader of Tachai Brigade, the national model for agricultural development. In September 1971, just about the time of Lin Piao's attempted escape and death, Ch'en criticized continued widespread opposition to agricultural mechanization: "Some people think [mechanization] is aimed only at reducing labor intensity and providing more leisure. . . . Others regard agricultural mechanization as an ordinary measure to save labor and increase production. . . . What [some people] worry about is that mechanization will lead to the loss of the revolutionary spirit of hard struggle."[160] Chen highlighted political and social issues:

> Mechanization is the Party's fundamental line in the rural areas for adhering to socialism and defeating capitalism. . . . [It is a] measure which consolidates the worker-peasant alliance, promotes socialist industrialization and reduces the differences between the workers and the peasants. Unless we implement Chairman Mao's revolutionary line mechanization will not necessarily bring about socialism and it may even lead to capitalism.[161]

159. Vail, "The Case for Rural Industry," pp. 40–43.

160. Ch'en Yung-kuei, "Ideological Revolutionization Leads to Farm Mechanization," *Peking Review*, no. 39 (Sept. 24), 1971, p. 13.

161. *Ibid*. Sometime in 1971 there was a national conference on farm mechanization to follow up the 1966 conference at Wuhan ("National Conference on Farm Mechanization Planned," NCNA Peking, Dec. 8, 1977). Perhaps these debates on mechanization surfaced at this conference.

Ch'en did not specify who the "some people" he criticized were, but Gerald Tannebaum, longtime resident of China, believed they were Lin Piao, Ch'en Po-ta, and others in opposition to Mao after the Cultural Revolution.[162] Their economic policies have never beeen specified in detail. Chou En-lai and others have described Lin's position to be that "the major contradiction in our country was . . . between the advanced socialist system and the backward productive forces of the society."[163] Exactly how Lin proposed to hasten the development of the productive forces of society has not been defined, but certain trends in the economy in 1970 suggest that he wanted to focus attention on the urban industrial system, on heavy industry, and perhaps on military procurement—all to the detriment of the rural sector.[164]

The general policy of emphasizing rural development came under attack from several points of view. The policy of keeping down prices of industrially produced agricultural inputs was criticized because it reduced the profits of factories—profits which could improve the welfare of factory workers and contribute to state income. Some factories resisted the call to place more emphasis on repair and maintenance services for agricultural machinery, as this cut into profitable production of new machinery. Likewise, some objected to demands that they contribute old machinery and the time of skilled workers and technicians to the development of rural local industries. Some manufacturers of consumer goods wanted to emphasize goods for the urban markets, as these were more profitable.[165]

The extent to which these views actually affected policy is difficult to determine, as detailed data on investments and production during this period are not available. There is some indication, however, that a substantial portion of the fruits of modernization during this period were used for military purposes. Estimates by the U.S. government of China's military expenses (both total expenses and procurement) shown in Figure 9.3 suggest that military expenses went up roughly 50 per cent from 1968 to 1971. In terms of absolute amounts, the total military expenses climbed

162. Gerald Tannebaum, *The Real Spirit of Tachai, Addendum* (New York: MSS Publications, 1974), p. 35.

163. Chou En-lai, "Report to the Tenth National Congress of the Communist Party of China," Aug. 24, 1973; available in *China Reconstructs*, special supplement, Nov. 1973.

164. Harding ("Political Trends in China," pp. 74–75) shares this view.

165. Leo Goodstadt, *China's Search for Plenty: The Economics of Mao Tse-tung* (New York: Weatherhill, 1973), pp. 203–207.

Figure 9.3. China's military expenses, 1964–73 (1964 = 100)

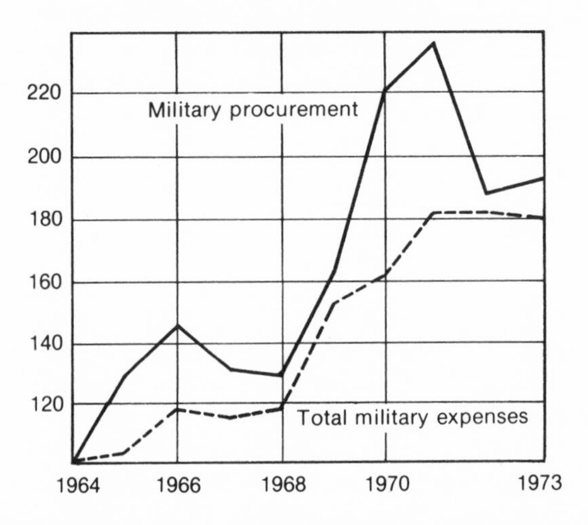

Sources:

Procurement: William Colby, "Allocation of Resources in the Soviet Union and China," *Hearings before the Subcommittee on Priorities and Economy in Government* (Washington, D.C.: U.S. Congress Joint Economic Committee, April 12, 1974), p. 77.

Expenses: *World Military Expenditures and Arms Trade, 1963–1973* (Washington, D.C.: U.S. Arms Control and Disarmament Agency, 1975), p. 27.

from about U.S. $8 billion in 1968 to about $14 billion in 1971. There can be no doubt that this expansion cut very deeply into the economy.

The main reason for the increase in military expenses at that time was of course the border clashes with the Soviet Union in 1969. Many undoubtedly felt that China was vulnerable to Soviet attack. The level of military expenses must have been subject to intense debate, and it seems likely that Lin Piao, in view of his military background and political base, would have favored increased military expansion. Although we lack data on the precise nature of military procurements and on the precise levels of production of agricultural machinery, it is likely that growth of agricultural mechanization was constrained by the increases in military procurement. Engines, for example, could have been diverted from tractor and pump factories and sent to tank and truck factories.

After the demise of Lin Piao (and the partial neutralization of the Soviet threat through adroit diplomacy, of which the visits of the U.S. ping-pong team, Henry Kissinger, and Richard Nixon were a part), it appears that China could again increase investments in (or reduce drain from) the rural sector. The precise extent of the investments is not yet known, but one American economist has estimated that the yearly state investment in agriculture went up from roughly ¥1 billion in the 1950's to about ¥5 billion in the early 1970's.[166] However, even the higher number is only 10–20 per cent of military expenditures.

In the months after Lin Piao's demise, price adjustments were made in industrial and agricultural commodities, so that more of the benefits of modernization would remain in the rural sector. In December 1971, the price of chemical fertilizers was reduced 9.7 per cent; of farm insecticides, 15 per cent; of kerosene, 20.8 per cent; of diesel oil, 9.7 per cent; and of diesel engines, harvesting machines, trucks, water pumps, and other items of farm machinery, an average of 15.7 per cent. At the same time, the procurement price for sugarcane went up 15.3 percent; and for peanuts, sesame, rapeseeds, and other oil-bearing crops and oils, an average 16.7 per cent.[167]

At the same time, the Central Committee issued a document on the question of distribution of income in the communes, which reaffirmed the general principles of commune organization that had characterized the system in the mid-1960's. The directive emphasized that proposals to increase drastically the level of investment in communes or the level of extraction from the state, as well as suggestions to level income between teams and individuals within the commune, were considered "ultra-leftist egalitarianism." It stated, "We should do everything possible to enable the peasants to raise their personal income year by year in normal years on the basis of increased production."[168]

The same directive also asked local Party leaders to combat sexual discrimination, which was widespread in the rural distribution system. "It is necessary to implement the principle of equal pay for equal work for both men and women."[169] It is not known whether Lin Piao supported sexist

166. Dwight Perkins, "China's Fourth Five-Year Plan," *Current Scene* 12:9 (Sept. 1974), 2.

167. China News Service, Peking, Dec. 26, 1971; NCNA Peking, Dec. 31, 1971; cited in "Increasing Priority for Chinese Agriculture," *Current Scene* 10:4 (April 1972), 16.

168. Central Committee Document No. 82, Dec. 26, 1971. Obtained by the Nationalist Chinese and released in *Central Daily News,* Aug. 18, 1972, and *Chung-kung Yen-chiu,* Sept. 1972. Translation in FBIS Republic of China Series, Aug. 25, 1972, pp. B1–5.

169. Some Pekinologists suggest that the programs to improve the status of women were directed at strengthening the political base of Chiang Ch'ing. Wang Ming, "The Fate of the 'Lonely Mong,' " *Mongolian Newspaper,* June 15–29, 1974; available in *Issues and Studies* 11:5 (May 1975), 113.

distribution policies or whether the struggle against Lin Piao took such priority that the political system could not focus on the question of sexism. However, it does seem that sexism was made a major issue after the fall of Lin Piao, and a widespread movement developed, of which this directive was a significant part.

To some extent agricultural mechanization has been used consciously to help women improve their relative position in Chinese society. There appears to be a policy that a very large portion of the tractor drivers should be young women. Statistics confirming this policy are unavailable, but my visual impressions while traveling in China indicate that a substantial portion of tractor drivers (especially the small tractors) are young women. The extent of propaganda on this subject is impressive; some Chinese currency carries pictures of women tractor drivers. Another way mechanization has reduced the burden on women has been through cutting the time and energy required for food preparation (still primarily a woman's task). The most important tools in this regard have been milling machines and fodder choppers, which lessen the effort required to feed pigs. Unfortunately, statistics are not available to show the diffusion of these machines. Women also find employment in small-scale industries, where they hold many jobs, including that of machine operator. Women do not appear to be given preferential opportunities in this employment; but they now have a chance to develop new skills, receive higher incomes, and attain a new type of status, based on their contribution to the national economy rather than on their performance of traditional female roles. They are creating new role-models for other young women who do not yet have jobs in industry. In this way rural industrialization is making a contribution to the changing of sex-linked roles in China.[170]

170. Perkins, *Rural Small-Scale Industry,* chapter 9, has a detailed analysis of the impact of rural industry on sexual roles.

10. Conclusion

From a mechanical engineering point of view, China's agricultural mechanization program has been competent and practical but not distinctive. China's tractors are very similar to—and in some cases are copies of—Soviet caterpillar tractors (which were originally copies of American tractors), English Massey-Ferguson tractors, and American tractors. Grain combines also appear to be based originally on Soviet, East European, and other designs. Smaller equipment (garden tractors, rice transplanters, small harvesters) is not very different from machinery used elsewhere in Asia, especially in Japan, although perhaps a bit heavier, more durable, and more reliable.[1] Horse-drawn implements (e.g., seed drills, plows) are undoubtedly designed by China's numerous craftsmen, but they look very similar to the analogous machinery on display in history of agriculture museums in the United States.

From an agroeconomic point of view, China's mechanization is not particularly unusual, either. In many places—China included—mechanization improves tillage, permitting crops to use fertilizer better and to reach higher yields. Mechanization of certain types reduces labor requirements for some tasks in agricultural production. In other applications, mechanization saves time and controls water more accurately, thus permitting increased multiple cropping. This can result in an overall increase in demand for labor.

The sociopolitical context shapes agricultural mechanization policy in several ways. First, because seasonal internal migration is virtually prohibited (except to suburban areas, where city workers and students may help in harvesting), mechanization is needed to break labor bottlenecks inherent in increasing multiple cropping. This is one of the reasons that China is paying special attention to mechanization.

1. Dwight Perkins, ed., *Rural Small-Scale Industry in the People's Republic of China* (Berkeley: University of California Press, 1977), p. 151.

Table 10.1. Rural price indexes (1952 = 100)

Year	Purchase price of farm and sideline products*	Price of industrial products sold in rural areas	Ratio of industrial price to agricultural price	Price of means of production sold in rural areas
1950[a]	82.2	91.2	111	
1952[a]	100	100	100	100
1954[a]	113.8	100.2	88	
1956[a]	116.6	100.4	86	
1958[a]	125.1	101.0	81	
1963[a]	154.7	114.3	74	
1964[b]			66	
1972[b]			60	
1973	160+[a]		approx. 57[c]	50[a]

Sources:

a. Dwight Perkins, "Constraints Influencing China's Agricultural Performance," in *China: A Reassessment of the Economy* (Washington, D.C.: U.S. Congress Joint Economic Committee, 1975), p. 362.

b. Jim Nickum, "A Collective Approach to Water Resource Development: The Chinese Commune System, 1962–72" (Ph.D. diss., University of California, Berkeley, 1974), p. 103.

c. Computed from data in Ernest P. Young, "Development, Chinese Style," *Understanding China Newsletter* 11:2 (March–April 1975), 3.

*The rise in purchase price of farm products probably reflects three factors: (1) the actual purchase price has gone up a few times, (2) probably a greater portion of state purchases are at a 30 percent bonus for grain sales in excess of targets that have been agreed upon by the production team and the state, and (3) there probably has been a shift in commodities to the more highly priced items, e.g., some regions have probably shifted from sales of millet and maize to wheat, others from wheat to rice.

Second, agricultural mechanization is being carried out in the context of a conscious policy to improve the rural standard of living, and to make it more equal to urban conditions. This policy has several dimensions. Prices of machinery (and of all other industrially manufactured products) are kept low, so that the beneficiaries of industrialization will be the rural population, not the owners, managers, and workers in the industrial sector[2] (see Table 10.1).

2. This policy is summed up by Carl Riskin: "China's current wage doctrine calls for maintaining relatively low wages for industrial workers, and limiting their rate of increase, in order to provide increasing surpluses for reinvestment, national defense, enlarging the sphere of collectively provided goods and services, and to some degree subsidizing the development of the countryside" ("Workers' Incentives in Chinese Industry," in *China: A Reassessment of the Economy* [Washington, D.C.: U.S. Congress Joint Economic Committee, 1975], p. 204.

The state further directs the benefits of industrialization to the rural sector by taxing the industrial sector far more heavily and then making investments in the countryside.

It should be noted that in China's socialist economy, government manipulation of farm and industrial prices can have a peculiarly powerful incentive effect. Not only the means of production but also the marketable surpluses are collectively owned, so virtually everyone in the rural sector benefits from higher state procurement prices for food grains and lower prices for means of production. This is in contrast to a system of private ownership, where very often only a small portion of farmers have net marketable surpluses and benefit from higher purchase prices, while a large portion of the rural inhabitants may in fact suffer from higher prices, as they are net purchasers of food grains. In other words, in China the ratio of prices between agricultural and industrial commodities is purely an urban-rural relationship, and not one of the wealthy farmers versus agricultural and industrial workers.[3]

One dimension of the standard of living is the character of work people do. Unmechanized farm labor is very grueling. Mechanizing farm work—including plowing, harvesting, irrigation, and grain processing—can contribute in a major way to reducing the toil and exposure to the elements of farm labor. At the same time, the policy of intermediate technology with rural factories generates increased nontraditional employment in the countryside. Together these elements are changing the character of employment in the countryside and making it more like that in the cities. The differences are still great, but a trend toward reduction of those differences has been established.

Another peculiarity of agricultural mechanization policy is that mechanization has been encouraged to strengthen the economic foundations of large-scale collective agriculture. In the 1950's, new plows were linked to collectivization; and in the late 1970's, in conjunction with the campaign to build Tachai-type counties, mechanization seemed to be related to plans to raise the level of accounting from the team to the brigade level. The larger unit was needed to purchase and utilize larger tractors; and the mechanization of agriculture would help equalize the conditions in teams, and thereby make it easier for teams to be merged. Another social reason for mechanization in the late 1970's was the hope

3. I am indebted to John Mellor and Dharm Narian for helping me understand this point.

that it would complement the family-planning program. It would be easier to convince peasants to keep families small if machinery were available to meet peak labor demands.[4]

To assure that the benefits of mechanization are widespread China has emphasized intermediate technology more than most other countries, including garden tractors and horse-drawn equipment. The Chinese have found that the benefits generated by large-scale ultramodern equipment tend to be concentrated in three ways. Only certain regions (especially North and Northeast China) have the flat lands and extensive agricultural systems in which large tractors can be used efficiently. Rice paddy regions of Central and South China can benefit from mechanical irrigation, grain processing, and transportation, but not from large tractors. Smaller-scale equipment, in contrast, can be used throughout China, in the rice-paddy regions and in the hilly regions where farm plots are small.

Second, high-technology equipment tends to require large-scale production facilities. This generally means that the benefits of industrial employment become concentrated in urban-industrial centers. Intermediate technology, on the other hand, is well suited to manufacture by rural industries which permit the industrial employment to be shared with rural residents.

Third, the large-scale modern equipment has been difficult to manage efficiently and to maintain. The Chinese have found tendencies toward the emergence of a new managerial elite which could capture a large share of the benefits of mechanization. Intermediate technology manufactured locally, in contrast, can be managed and maintained more easily with local personnel.

Some resistance to intermediate technology has been encountered. For reasons of status and ideology, some local officials have wanted fully modern equipment, and they have feared that if they invest in intermediate technology, they will have reduced opportunities to obtain modern machinery. In theory, intermediate technology offers the opportunity to generate capital and train people in relevant skills, thus facilitating the procurement of modern equipment at a later time.

The emphasis on intermediate technology should not be overstressed. China has, at the same time, developed large-scale, modern agricultural

4. Chin Chi-chu, "Farm Mechanization in Wusih County," *Peking Review*, no. 31 (July 29), 1977, pp. 20–24, and no. 33 (Aug. 12), 1977, pp. 36–39.

machinery where it is required. The Chinese policy in this regard, as in others, is pragmatically to "walk on two legs."

The most striking aspect of agricultural mechanization is that its benefits are shared widely throughout the rural community because of the institution of collective ownership. It is in this dimension that China is very different from most other places. In countries with private agriculture, the analyses of mechanization revolve around the question of landlord-tenant relations. Will mechanization provide the technical-economic basis for landowners to consolidate and enlarge their holdings? Will they push tenants off their holdings? Will mechanization result in a greater polarization of wealth in the countryside? Will it result in increased unemployment and urban migration?

Because of collective ownership in China, everyone benefits. If mechanization reduces costs and increases the profitability of farming, then the value of the work points of everyone will rise. If mechanization reduces labor requirements, then everyone will benefit from less labor intensity; the burdens of unemployment will not fall on a few.

Collective ownership of machinery seems to have been efficient from an economic point of view, especially after enough people at the commune level were trained to operate and maintain machinery. Unlike state or bureaucratic ownership systems, the owners of machines in China actually use the machines and profit or loose because of them. Thus, practical incentives facilitate the development of sensible, pragmatic machines, well adapted to local needs. A centralized bureaucracy has not been necessary for efficiency, and in fact has been counterproductive.

While collective ownership of machines has been a useful counterweight to bureaucratic growth, it is not by itself a completely adequate safeguard. Some Chinese would be quick to point out that even though kolkhozy now own machinery in the Soviet Union, bureaucratic elitism permeates both local collective organizations and higher levels. Political consciousness, some Chinese would argue, is more important and must supplement structural reforms. In the long run, they might admit, there is no guarantee that even this can block the pressures towards bureaucratization.

What can be said in a broad sense about the political system that has selected these policies? Is it subject to great pressures toward the irrational? Does radical ideology stand in the way of rational economic policies? Certainly at some times it has. The almost desperate attempts to find farm tools that would consolidate the collective economy in the 1950's (the two-wheel two-blade plows, the rice transplanters, the cable-drawn plows) undoubtedly resulted in much waste. But one can not help but be impressed by the suitability of Chinese agricultural mechanization policy since the early 1960's. It has recognized the importance of complementary inputs of chemicals, seeds, and better water control. It has succeeded in mobilizing both capital and labor for flood control, drainage, and improvement projects. It has recognized intermediate technology as a way of assuring that benefits of modernization would be widely spread geographically. It has found a way of supporting modern technology without strengthening a class of technical managers, who could become a new exploitative elite. It has assured that virtually everyone in a village will benefit from modernization, and that unemployment will not threaten the livelihood of a significant group. In a broad sense, China has drawn on global experience for both technology (e.g., fertilizer factories) and ideas (e.g., the need to be cautious in collectivizing agriculture). At the same time, China has developed policies which are geared to specific local requirements in terms of culture, economic systems, and natural ecology.

These policy choices have been made consistently, despite intense political conflict at the highest levels, despite numerous purges, and despite the apparent disarray of high-level political institutions. Indeed, many in China would argue that it is precisely the constant struggle and disarray which has been essential in weakening the bureaucracy and urban industrial interests and allowing rural interests to prevail.

Ideology, which is often accused of inspiring irrational, rigid, unpragmatic policies, may very well be the crucial reason for this consistent adoption of sensible policies. The notions of "serve the people," "reduce the three differences (between city and countryside, industrial work and farm work, and mental and manual labor)," "mass line"—are cables running through Chinese Communist ideology, binding policy to programs that have increasing and widely distributed benefits. The manner in which these slogans are transformed into concrete policies is, of course, the core of Chinese politics. Perpetual political struggle and education are integral parts of this process. It has never been fully described, and may never be because of inadequate data.

Even if the precise mechanisms of the political system cannot be understood and described fully, a crucial conclusion can be drawn. In developing sensible policies, ideology and politics have been important positive forces. The notion that development can proceed by itself without political leadership is certainly challenged by China's experience.

China is groping toward a way of making modern technology serve everyone. No large segments of the population are left out of development, nor are certain segments seizing most of the benefits. Deeply embedded in the culture there remain pressures toward the emergence of a new ruling class which draws its power from a bureaucracy which manages the economy; China seems to be keeping under control the bureaucratic tendencies which Western social science presumes to be synonymous with modernization, but constant political struggle must be expected, and the political resources of the bureaucracy must not be underestimated. China is demonstrating on a massive scale that modernization need not follow any rigid path. Within limitations, there is ample room for political choice. Perhaps this is the important lesson to be learned from China's experience: that the processes of modernization and technological development—in which all peoples of the globe are involved—are not subject to immutable laws of heaven or earth, but are shaped and reshaped through human-made political institutions.

Appendix 1. Semimechanized and Improved Tools Produced, 1950–57 (all figures in parentheses are estimates)

Year	2-wheel plows		Walking plows		Sprayer-dusters		Waterwheels	
	Produced*	Cumulative	Produced	Cumulative	Produced	Cumulative	Produced	Cumulative
1950	166[a]		594[a]				67,270[a]	
1951	582[a]	748	5,717[a]	6,311			96,095[a]	163,365
1952	5,000[b]	5,748	237,368[a]	243,679	164,047[a]		222,718[a]	386,083
1953	3,000[b]	8,748	178,033[a]	421,712	197,988[a]	362,045	174,562[a]	560,645
1954	60,000[b]	68,748	224,242[a]	645,954	309,974[a]	672,019	117,816[a]	678,461
1955	523,000[b]	591,748	304,757[a]	950,711	429,215[a]	1,101,234	164,444[a]	842,905
1956	1,793,000[b]	2,384,748	1,022,464[a]	1,973,175	1,179,952[a]	2,281,186	619,436[a]	1,462,341
1957	1,800,000[b]	4,184,748	96,945[a]	2,070,120	540,526[a]	2,821,712	247,222[a]	1,709,563

Year	Threshers		Rollers		Reapers		Rakes	
	Produced	Cumulative	Produced	Cumulative	Produced	Cumulative	Produced	Cumulative
1950								
1951								
1952	(50,000)		(15,000)					
1953	(50,000)	(100,000)	(15,000)	30,000[d]			(10,000)	
1954	(75,000)	(175,000)		(28,000)	(5,000)		(10,000)	(20,000)
1955	(75,000)	(250,000)		(28,000)	(5,000)	(10,000)	(20,000)	(40,000)
1956	(93,000)	343,000[c]		28,000[e]	(5,000)	(15,000)	(23,000)	63,000[e]
1957		318,000[f]	(8,000)	36,000[f]	(5,000)	20,000[f]	(13,000)	76,000[f]

Year	Harrows		Seeders		Weeding plows		Processing implements	
	Produced	Cumulative	Produced	Cumulative	Produced	Cumulative	Produced	Cumulative
1950								
1951								
1952							(50,000)	
1953	(11,000)						(100,000)	150,000[d]
1954	(15,000)	(25,000)	(10,000)				(30,000)	(180,000)
1955	(15,000)	(41,000)	(13,000)	(23,000)	(50,000)		(20,000)	(200,000)
1956	(22,000)	63,000[c]	(25,000)	48,000[c,b]	(80,000)	130,000[c]	(30,000)	(230,000)
1957	(10,000)	(73,000)	(8,000)	56,000[e]	(189,000)	319,000[e]	(28,000)	258,000[e]

Sources:

a. "History of Struggle," SCMM 633, pp. 6–7. This article gives figures for tools "introduced," which may be lower than the number of tools actually produced.

b. Kang Chao, *Agricultural Production in Communist China, 1949–1965* (Madison: University of Wisconsin Press, 1970), p. 102.

c. Chi Chung-wei, "China's Industry to Support Agricultural Production," *Ching-chi Yen-chiu,* no. 2 (Feb. 17), 1958 (ECMM 127, p. 50).

d. *Hsin-hua Yueh-pao,* 1955, no. 4, p. 134; cited by Chao, p. 101.

e. *Chung-kuo Nung-pao,* 1958, no. 3, p. 8; cited by Chao, p. 101.

*Not all the 2-wheel plows that were produced were actually "introduced." Only 1,625,000 of the over 4,000,000 produced were reported to have been "introduced" ("History of Struggle," SCMM 633, pp. 6–7).

Appendix 2. Production of Semimechanized Tools, 1960–63

Statistics on production of semimechanized tools from 1960 to 1963 are unclear and incomplete. Chun Wen, in an article in *Workers' Daily* in September 1963, claimed that 30 million semimechanized small farm implements had been supplied from 1960 to 1962.[1] This figure seems either to be inflated or to include smaller tools not usually considered to be "semimechanized," because in October 1964, Li Chi-huan, vice minister of Agricultural Machinery, claimed that 20 million pieces of semimechanized and improved tools had been produced from 1960 to 1964.[2] This figure seems acceptable because it is in line with a claim at a 1966 national exhibition that in 1965 there were 2.78 times as many medium-sized semimechanized farm tools as in 1956.[3] My estimates are that by 1956, 8–9 million such tools had been supplied, and by 1965, 22–25 million such tools were in use. This agrees with the claim of 20 million tools supplied from 1960 to 1964, and would mean that 5–8 million tools that had been supplied were not in use—either due to normal depreciation or because they were not technically satisfactory, as for instance the new plows introduced in 1956 and the 2.3 million transplanters popularized in 1960.

A precise breakdown of the 20 million semimechanized tools supplied from 1960 to 1964 is unavailable, but in the years 1960 and 1961, 10 million wheelbarrows and 150,000 carts were produced;[4] in 1960, 2.3

1. Chun Wen, "New Developments in the Production of Agricultural Machinery," *Kung-jen Jih-pao*, Sept. 24, 1963 (SCMP 3089, p. 16).

2. Li Chi-huan, "Further Develop Semi-Mechanized Farm Tools to Provide Better Assistance to Farm Production," *Jen-min Jih-pao*, Oct. 18, 1964 (SCMP 3332, p. 1).

3. "Paving a Way to Agricultural Mechanization in China—Visit to Agricultural Mechanization Hall of the National Agricultural Exhibition," *ibid.*, April 13, 1966 (SCMP 3684, p. 9).

4. "New Achievements by China's Agricultural Machine Industry," *Shih-shih Shou-ts'e*, no. 3–4 (Feb. 17), 1962 (SCMM 315, p. 21).

million rice transplanters were produced.[5] If we assume the wheelbarrows and carts were produced equally in both years, then at least 7.37 million semimechanized implements were produced in 1960 and at least 5.07 million were produced in 1961. In 1962–63, fewer than 7.7 million (i.e., less than 3.77 million each year) were supplied.

5. "Chinese Peasants Make New Tools for Rice Fields," NCNA Peking, July 20, 1960 (SCMP 2304, p. 4).

Selected Bibliography

Primary Sources

"Basic Summary of the Work of Breeding Seedlings of Low-stalk Strains of Rice in Kwangtung." *Kuang-tung Nung-yeh K'o-hsüeh,* no. 1 (Feb. 20), 1966. JPRS 36,710, p. 2.

"CCPCC and State Council Issue Directive on Movement for Improvement of Agricultural Implements in Rural Areas." NCNA Peking, July 14, 1958. SCMP 1819, p. 5.

Central–South China Bureau of the Chinese Communist Party Central Committee, Field Team of the Economic Commission. "Operate Hsien-Level Agricultural Machinery Plants Properly to Support Agricultural Production." *Chung-kuo Nung-yeh Chi-hsieh,* no. 4 (April), 1965, pp. 7–8. JPRS 42,484, p. 13.

Chang Ch'ing-t'ai. "On the Question of Strengthening the Operation and Management of Tractor Stations," excerpts from a speech delivered at Liaoning Provincial Conference on Tractor Station Work. *Chung-kuo Nung-pao,* no. 11 (Nov. 10), 1963. SCMM 400, pp. 22–33.

Chang Feng-shih. "Produce Suitable Agricultural Machinery to Support Agriculture." *Kung-jen Jih-pao,* Jan. 18, 1962. JPRS 12,909, p. 11.

Chao Hsüeh. "The Problem of Agricultural Mechanization in China." *Chi-hua Ching-chi,* no. 4 (April 9), 1957. ECMM 87, p. 10.

——. "The Problem of Farm Mechanization in China." *Jen-min Jih-pao,* Oct. 24, 25, 1957. SCMP 1662, p. 14.

——. "Several Current Problems of Agricultural Mechanization." *Ta Kung Pao,* Peking, May 15, 1961. SCMP 2515, p. 23.

Ch'en Yung-kuei. "Ideological Revolutionization Leads to Farm Mechanization." *Peking Review,* no. 39 (Sept. 24), 1971, p. 13.

Chiang Wen-hsien. "How the Tractor Station of T'aiku Hsien Serves Agricultural Production." *Ching-chi Yen-chiu,* no. 4 (April), 1965. SCMM 472, p. 18.

Chieh-fang Jih-pao editor, with editors of *Wen-hui Pao, Party Branch Life.* "Two Diametrically Opposed Lines in Building the Economy." *Jen-min Jih-pao,* Aug. 25, 1967. SCMP 4012.

China Institute of Research in Agricultural Mechanization, Capital Workers' Congress, East Is Red Commune. "Last-Ditch Struggle That Courts Self-Destruction—Denouncing the Towering Crimes of China's Khrushchev in Opposing Chairman Mao's Wise Decision." *Nung-yeh Chi-hsieh Chi-shu,* no. 5 (Aug. 8), 1967. SCMM 613, pp. 24–28.

China Scientific Research Institute in Agricultural Mechanization. ''Resolutely Criticize and Repudiate China's Khrushchev for Committing the Crime of Undermining the Cause of Agricultural Mechanization.'' *Nung-yeh Chi-hsieh Chi-shu,* no. 2–3 (May 23), 1967. SCMM 590, pp. 20–25.

''China's Agricultural Machine Industry Must Take the Road Indicated by Chairman Mao.'' See: Eighth Ministry of Machine Building, Research Office of the United Headquarters, Support the Peasants and Soldiers Fighting Detachment.

Ching Hung. ''The Plot of the Top Ambitionist to Operate 'Trusts' on a Large Scale Must Be Thoroughly Exposed.'' *Kuang-ming Jih-pao,* May 9, 1967. SCMP 3948, p. 6.

Ching Tan. ''Machinery Especially for Chinese Farms.'' *China Reconstructs,* Aug. 1964, p. 6.

''Completely Settle the Heinous Crimes.'' See: Peking Agricultural Machinery College, East Is Red Commune, Criticism and Repudiation Office.

Eighth Ministry of Machine Building, Agricultural Machinery Management Bureau, Red Machine Troops. ''China's Khrushchev's Crime of Thwarting the Farm Tool Renovation Movement Must Be Thoroughly Reckoned With.'' *Nung-yeh Chi-hsieh Chi-shu,* no. 3, 1968. SCMM 624, pp. 8–11.

Eighth Ministry of Machine Building, Agricultural Machinery Management Bureau, Revolutionary Great Alliance Committee, Mass Criticism Unit. ''History of Struggle between the Two Lines (on China's Farm Machinery Front).'' *Nung-yeh Chi-hsieh Chi-shu,* no. 9, 1968. SCMM 633, pp. 1–34.

——. ''T'an Chen-lin's Crime of Sabotage against Farm Mechanization Must Be Reckoned With.'' *Nung-yeh Chi-hsieh Chi-shu,* no. 6 (June 8), 1968. SCMM 624, pp. 1–7.

Eighth Ministry of Machine Building, Research Office of the United Headquarters, Support the Peasants and Soldiers Fighting Detachment. ''China's Agricultural Machine Industry Must Take the Road Indicated by Chairman Mao.'' *Nung-yeh Chi-hsieh Chi-shu,* no. 9, 1968. SCMM 644, pp. 20–28.

Eighth Ministry of Machine Building, Revolutionary Great Alliance Headquarters and Revolutionary Great Criticism and Repudiation Group of Organizations. ''Two Diametrically Opposite Lines in Agricultural Mechanization.'' *Nung-yeh Chi-hsieh Chi-shu,* no. 9, 1968. SCMM 633, pp. 35–45.

Eighth Ministry of Machine Building, United Committee of the Revolutionary Rebels, ''57'' United Detachment. ''Wipe out State Monopoly and Promote Mechanization on the Basis of Self-Reliance in a Big Way.'' *Nung-yeh Chi-hsieh Chi-shu,* no. 6 (Sept. 18), 1967. SCMM 610, pp. 10–16.

Feng Chih-kuo. ''How to Solve the Problem of Labor Shortage on the Agricultural Front.'' *Ching-chi Yen-chiu,* no. 3 (March 17), 1959. ECMM 167, p. 18.

Heilungkiang Communist Party, Paichuan County Committee and Nunkiang District Committee, and Heilungkiang Branch of NCNA. ''The Universal Establishment of People's Communes Accelerates Mechanization—An Investigation into Hsing-nung People's Commune in Paichuan County, Heilungkiang Province.'' *Jen-min Jih-pao,* Dec. 4, 1959. SCMP 2160, p. 10.

Heilungkiang Department of Agricultural Mechanization, Committee Taking over the Control. ''Let the Radiance of Mao Tse-tung's Thought for ever Shine over the Road of Agricultural Mechanization." *Nung-yeh Chi-hsieh Chi-shu,* no. 4 (July 8), 1967. SCMM 600, pp. 1–7.

Heilungkiang Department of Agricultural Mechanization, Mao Tse-tung Thought Red Guards and the Red Rebel Regiment. "Thoroughly Wipe out the Pernicious Influence of the Bourgeois Reactionary Line, Let the Great Red Banner of Mao Tse-tung's Thought Be Planted all over the Agricultural Mechanization Front." *Nung-yeh Chi-hsieh Chi-shu,* no. 2–3 (May 23), 1967. SCMM 588, pp. 10–13.

Heilungkiang, Nunchiang Country, Take-over Committee of the Agricultural Machine Station. "Thoroughly Eliminate the Pernicious Influence of Material Incentive." *Nung-yeh Chi-hsieh Chi-shu,* no. 4 (July), 1967. SCMM 605, p. 30.

Heilungkiang, Work Group of CCP Nunkiang District Committee, CCP Paichuan County Committee of Heilungkiang, and Heilungkiang Branch of the New China News Agency. "The Universal Establishment of People's Communes Accelerates Mechanization—An Investigation into Hsingnung People's Commune in Paichuan Hsien, Heilungkiang Province." *Jen-min Jih-pao,* Dec. 4, 1959. SCMP 2160, p. 10.

"History of Struggle." See: Eighth Ministry of Machine Building, Agricultural Machinery Management Bureau, Revolutionary Great Alliance Committee, Mass Criticism Unit.

Hopei, Chengting County Agricultural Machine Work Station. "March with Big Strides on the Revolutionary Road—How Chengting Hsien Station's Farm Machines and Tools Were Sent Down." *Nung-yeh Chi-hsieh Chi-shu,* no. 11 (Nov. 8), 1968. SCMM 643, pp. 8–12.

How the Soviet Revisionists Carry out All-Round Restoration of Capitalism in the U.S.S.R. Peking: Foreign Languages Press, 1968.

Hsiang Nan. "Certain Problems of Agricultural Mechanization." *Jen-min Jih-pao,* Dec. 22, 1962. SCMP 2900, p. 1.

——. "An Inspection Report on the Mechanization of Our Agriculture." *Chung-kuo Nung-yeh Chi-hsieh,* May 1965. JPRS 33,691, pp. 14–29. Also available as "Stable and High Yields and Agricultural Mechanization," *Jen-min Jih-pao,* March 22, 1965, SCMP 3436, pp. 10–16, and "Agricultural Mechanization Can Be Achieved with Good and Fast Results," *Jen-min Jih-pao,* July 6, 1965, SCMP 3518, pp. 3–8, in edited form.

Hsü Pin-chou. "Struggle for the Realization of Agricultural Mechanization." *Kung-jen Jih-pao,* Oct. 10, 1964. SCMP 3329, p. 11.

Huang Ching. "The Problem of Farm Mechanization in China." *Jen-min Jih-pao,* Oct. 24–25, 1957. SCMP 1662, p. 14. Also in *Chi-hsieh Kung-yeh,* Nov. 1957, ECMM 120, p. 34, under the title "On Agricultural Mechanization in China."

——. "Simultaneous Development of Industry and Agriculture and the Question of Agricultural Mechanization." *Hsüeh-hsi,* no. 2 (Jan. 18), 1958. ECMM 128, p. 54.

Hunan, Office of the Department of Hydroelectricity, Water Wheel Pump Office, September 17 Red Fighting Regiment. "Completely Settle Scores with the Crimes Committed by T'an Chen-lin and His Like in the Construction of Water

Wheel Pumps.'' *Nung-yeh Chi-hsieh Chi-shu,* no. 7 (July 8), 1968. SCMM 624, p. 12.

Hung Yen Ping [Red Research Troops]. ''A Newborn Thing Is Undefeatable, Exposure and Criticism of Big Traitor T'an Chen-lin's Crime of Undermining the Experimentation of Power-Operated Plows.'' *Nung-yeh Chi-hsieh Chi-shu,* no. 7 (July 8), 1968. SCMM 624, pp. 17–19.

Hupeh, O-ch'eng County, Production Brigade No. 2, Hsukuang Commune. ''The Question of Mechanization by the Collective.'' *Nung-yeh Chi-hsieh Chi-shu,* no. 9 (Sept. 1968), supplement. SCMM 644, pp. 13–19.

''It Is Good for Tractors to Revert to Chairman Mao's Revolutionary Line—Report on Investigation of Change in Management of Tractors by the Collective in Lank'ou County, Honan.'' *Nung-yeh Chi-hsieh Chi-shu,* no. 10 (Oct. 8), 1968. SCMM 643, pp. 17–26.

''Joint Directive Issued on Farm Tools.'' NCNA, Peking, March 15, 1958. SCMP 1758, p. 11.

Kiangsu, Hsinyi County, Agricultural Machinery Corporation, Red Workers' Combat Detachment Defending the Thought of Mao Tse-tung. ''Refute Material Incentive and Render It Repugnant.'' *Nung-yeh Chi-hsieh Chi-shu,* no. 4 (July 8), 1967. SCMM 605, p. 27.

Kuo Kuo-sheng. ''Development of Potentials of Plowing by Tractors as Seen from the Case of Two Tractor Stations.'' *Kung-jen Jih-pao,* May 17, 1962. SCMP 2752, p. 15.

Kwangtung, Huaichi County, Agricultural Mechanization Office of Lengk'eng Commune. ''Initial Experience of Lengk'eng Commune in Promoting Agricultural Mechanization on a Large Scale.'' *Nung-yeh Chi-hsieh Chi-shu,* no. 4 (July 8), 1967. SCMM 600, pp. 8–16.

Kwangtung Provincial Land Reform Committee, Investigation Section. ''Mass Reaction and Problems after Land Reform.'' *Nan-fang Jih-pao,* Feb. 19, 1953. SCMP 527, p. 21.

Lao T'ung-ping. ''Thoroughly Bury Foreign Slave Philosophy.'' *Kuang-ming Jih-pao,* March 22, 1969. SCMP 4395, pp. 6–7.

''Last-Ditch Struggle.'' See: China Institute of Research in Agricultural Mechanization, Capital Workers' Congress, East Is Red Commune.

''Let the Radiance of Mao Tse-tung's Thought.'' See: Heilungkiang Department of Agricultural Mechanization, Committee Taking over the Control.

Li Chi-hsin. *Nung Chi Chan* [Agricultural Machinery Stations]. Shanghai: Commercial Press, 1950.

Li Chien-pai. ''Questions of Agricultural Mechanization.'' *Hsüeh-hsi,* no. 10–11 (May 31), 1958. ECMM 140, pp. 37–41.

Li Ch'ing-yü. ''Realize Step-by-Step Agricultural Mechanization through Tool Innovation.'' *Chung-kuo Nung-pao,* Oct. 8, 1959. ECMM 194, p. 33.

——. ''A Summary of the Conference on Agricultural Mechanization and Electrification.'' *Nung-yeh Chi-hsieh,* no. 1 (Jan. 15), 1959. ECMM 161, p. 12.

Li She-nan. ''A Few Problems Regarding the Lowering of Operational Costs of Agricultural Mechanization.'' *Jen-min Jih-pao,* June 30, 1964. SCMP 3272, p. 1.

Liang Hsiu-feng. "Preliminary Enquiry into the Center, Step and Keypoint of Agro-technical Reform in Our Country." *Ching-chi Yen-chiu,* no. 9 (Sept. 17), 1963. SCMM 389, p. 36.

Liao Nung-ko. "Knock Down China's Khrushchev and Completely Discredit by Criticism His Line of State Monopoly." *Nung-yeh Chi-hsieh Chi-shu,* no. 2, 1968. SCMM 620, pp. 5–9.

——. "Material Incentive Is a Poison to the Revolutionary Masses." *Nung-yeh Chi-hsieh Chi-shu,* no. 6 (June 8), 1968. SCMM 620, p. 29.

Liu Chih-cheng, Ho Kuei-t'ing, and Hsu Hsing. "The Relations between the Four Transformations and Economic Effects." *Ching-chi Yen-chiu,* no. 2, 1964. SCMM 424, pp. 1–15.

Liu Chung-huang. "On the Technical Transformation of China's Agriculture." *Jen-min Jih-pao,* Aug. 26, 1960. SCMP 2333, p. 1.

Liu Jih-hsin. "Exploration of a Few Problems Concerning Mechanization of Our Agriculture." *Jen-min Jih-pao,* June 20, 1963. SCMP 3021, p. 1.

Mao Tse-tung. *Miscellany of Mao Tse-tung Thought.* Arlington, Virginia: Joint Publications Research Service, 1974.

——. *Selected Readings from the Works of Mao Tsetung.* Peking: Foreign Languages Press, 1971.

——. "Sixty Work Methods [Draft]," January 31, 1958. *Long Live Mao Tse-tung Thought,* CB 892, p. 13.

"March with Big Strides." See: Hopei, Chengting County Agricultural Machine Work Station.

Miscellany. See: Mao Tse-tung.

Northeast Bureau of the Chinese Communist Party Central Committee. "Decisions on Strengthening Work of State Farms, September 1952." *Jen-min Jih-pao,* Dec. 10, 1952. SCMP 485, p. 16.

Ou-yang Ch'in. "Overall Planning and 'Walking on Two Legs' Essential to Agricultural Mechanization." *Hung-ch'i,* no. 7 (April 1), 1960. ECMM 209, pp. 7–19.

——. "Questions Concerning Agricultural Mechanization." *Jen-min Jih-pao,* Dec. 21, 1959. SCMP 2174, p. 23.

Peking Agricultural Machinery College, East Is Red Commune, Criticism and Repudiation Office. "Completely Settle the Heinous Crimes of China's Khrushchev and Company in Undermining Agricultural Mechanization." *Nung-yeh Chi-hsieh Chi-shu,* no. 5 (Aug. 8), 1967. SCMM 610, pp. 17–32.

——. "The P'eng Chen Counter-Revolutionary Revisionist Clique's Crime Is Most Heinous." *Nung-yeh Chi-hsieh Chi-shu,* no. 6 (Sept. 18), 1967. SCMM 610, pp. 5–9.

Peking Institute of Agricultural Mechanization, Capital Red Guard Congress, "East Is Red" Commune. "The Reactionary Nature of China's Khrushchev in Promoting the Trust." *Nung-yeh Chi-hsieh Chi-shu,* no. 5 (Aug. 8), 1967. SCMM 613, p. 20.

"P'eng Chen Counter-Revolutionary Revisionist Clique's Crime." See: Peking Agricultural Machinery College, East Is Red Commune, Criticism and Repudiation Office.

People's Daily editor. " 'Demonstration Farms' are Main Centers through which Agricultural Science May Serve Production." *Jen-min Jih-pao,* Oct. 23, 1964. SCMP 3338, p. 14.

——. "Editor's Note on the Postulation by CCP Hupeh Provincial Committee Concerning Gradual Realization of Agricultural Mechanization." *Jen-min Jih-pao,* April 9, 1966. SCMP 3679, p. 14.

——. "A Great Campaign for Improving Tools." *Jen-min Jih-pao* editorial, Aug. 21, 1958. SCMP 1845, p. 21.

——. "Launch an All-People Campaign for Mechanization and Semi-Mechanization of Manual Labor." *Jen-min Jih-pao,* Feb. 25, 1960. SCMP 2212, p. 8.

——. "Management of Agricultural Machinery Should Better Serve Agricultural Production." *Jen-min Jih-pao,* Aug. 31, 1965. SCMP 3543, p. 9.

——. "New Development of Tools Innovation in Rural Areas." *Jen-min Jih-pao,* Jan. 13, 1960. SCMP 2180, p. 6.

——. "Pay Attention to Maintenance, Repair, and Production of Medium-sized Farm Implements." *Jen-min Jih-pao,* May 20, 1962. SCMP 2750, p. 11.

——. "Properly Manage and Make Better Use of Farm Machinery." *Jen-min Jih-pao,* June 11, 1961. SCMP 2526, p. 11.

——. "Raise Agricultural Science to a New Level." *Jen-min Jih-pao,* April 6, 1963. SCMP 2964, p. 6.

——. "Run Tractor Stations Truly Well." *Jen-min Jih-pao,* Nov. 26, 1963. SCMP 3115, pp. 9–11.

——. "Running Enterprises in Line with Mao Tse-tung's Thinking." *Jen-min Jih-pao,* April 3, 1966. Also *Peking Review,* no. 16 (April 15), 1966, p. 11.

——. "Speed up the Improvement of Farm Tools." *Jen-min Jih-pao,* Jan. 21, 1959. SCMP 1959, p. 20.

——. "The Spirit of Self-Reliance Must Be Upheld in Farm Implement Innovation Movement." *Jen-min Jih-pao,* Oct. 15, 1964. SCMP 3329, p. 14.

Planned Economy editor. "Why Does the Demand for Double-Wheel and Double-Blade Plows Drop and Why Is Their Production Suspended?" *Chi-hua Ching-chi,* no. 9 (Sept. 23), 1956. ECMM 56, p. 24.

"Postulation by the CCP Hupeh Provincial Committee Concerning Gradual Realization of Agricultural Mechanization." *Yang-ch'eng Wan-pao,* Canton, April 4, 1966. SCMP 3679, p. 8.

"Reactionary Nature of China's Khrushchev." See: Peking Institute of Agricultural Mechanization, Capital Red Guard Congress, East Is Red Commune.

"Refute Material Incentive and Render It Repugnant." See: Kiangsu, Hsinyi County, Agricultural Machinery Corporation, Red Workers' Combat Detachment Defending the Thought of Mao Tse-tung.

"Rely on the Masses." See: Shansi, Ta-t'ung City Farm Machine Station of Chaochia hsiao-ts'un Commune.

Shang Chin-lung and Ma Ching-p'o. "Fifteen Years of Agricultural Mechanization on State Farms." *Nung-yeh Chi-hsieh Chi-shu,* no. 11 (Nov.) 1964. SCMM 451, pp. 6–9.

Shanghai Municipal Bureau of Agriculture, Office of Agricultural Machinery Building, Rebel Team. "China's Khrushchev and His Agents in Shanghai Can

Hardly Be Acquitted of Their Crime in Sabotaging Agricultural Mechanization." *Nung-yeh Chi-hsieh Chi-shu,* no. 6 (Sept.), 1967. SCMM 611, p. 30.

Shansi, Jui-ch'eng County Revolutionary Committee. "Use Mao Tse-tung's Thought to Direct the Work of Sending down the Farm Machines and Tools of State Stations." *Nung-yeh Chi-hsieh Chi-shu,* no. 11 (Nov. 8), 1968. SCMM 643, pp. 3–7.

Shansi, Ta-t'ung City Farm Machine Station of Chaochia hsiao-ts'un Commune. "Rely on the Masses to Run the Farm Machine Station Democratically." *Nung-yen Chi-hsieh Chi-shu,* no. 9 (Sept. 1968), supplement. SCMM 644, pp. 7–12.

Shensi Bureau of Agricultural Machinery, Revolutionary Leading Group. "T'an Chen-lin Cannot Shun His Criminal Responsibility for Closely Following China's Khrushchev in Promoting Capitalist Trust." *Nung-yeh Chi-hsieh Chi-shu,* no. 9, 1968. SCMM 644, pp. 35–38.

"Struggle between Two Lines as Viewed from the Reversals in Production of Feng-shou Tractors." *Nung-yeh Chi-hsieh Chi-shu,* no. 6, 1967. SCMM 609, p. 14.

Su Hsing. "The Struggle between Socialist and Capitalist Roads in China after Land Reform." *Ching-chi Yen-chiu,* nos. 7–9 (July–Sept.), 1965. SCMM 495, pp. 1–18; SCMM 498, pp. 1–16; SCMM 499, pp. 19–33.

"Suggestions Concerning Strengthening the Management of Agricultural Machinery by People's Communes, Revised Draft." *Nung-yeh Chi-hsieh,* no. 1, 1959, p. 20.

"Summing-up Report on the National Conference of Tractor Station Masters (Excerpts)." *Chung-kuo Nung-pao,* no. 5 (Nov.), 1958. ECMM 147, pp. 16–23.

Sung Wei-ching. "Accelerate the Agro-technical Reform, Deepen the Farming Tools Innovation Movement." *Hung-ch'i,* no. 15 (Aug. 1), 1960. SCMM 228, p. 18.

T'an Chen-lin. "Report on Program for Agricultural Development." NCNA Peking, April 6, 1960. CB 616, p. 22.

"T'an Chen-lin Cannot Shun His Criminal Responsibility." See: Shensi Bureau of Agricultural Machinery, Revolutionary Leading Group.

"T'an Chen-lin's Crime of Sabotage." See: Eighth Ministry of Machine Building, Agricultural Machinery Management Bureau, Revolutionary Great Alliance Committee, Mass Criticism Unit.

T'ao Chu. "The Tasks of the Party Organization in South China during the State's Five-Year Construction Plan." *Nan-fang Jih-pao,* Oct. 31, 1953. SCMP 703 supplement, p. iv.

"Telephone Conferences Held on Tools Improvement: Instruction Given by Chairman Mao." *Jen-min Jih-Pao,* Aug. 21, 1958. SCMP 1845, p. 20.

Teng Chieh. "Bring the Role of Handicrafts into Full Play, Serve Agricultural Production Better." *Jen-min Jih-pao,* Jan. 4, 1966. SCMP 3619, p. 7.

"Thoroughly Eliminate the Pernicious Influence of Material Incentive." See: Heilungkiang, Nunchiang County, Take-over Committee of the Agricultural Machine Station.

Ting Li-ch'u. "Exploration into the Rational Management of the State Farm: Investigation and Study of the Problem of the Management of State Farms in Heilungkiang.'' *Ching-chi Yen-chiu,* no. 12, 1963. SCMM 403, p. 12.

T'o-la-chi Kuo-tsao [Structure of Tractors]. Peking: 1973. Available in *Translations on People's Republic of China,* no. 287, p. 18. JPRS 63,091, pp. 14–36.

Tso Hu. ''Some Problems of Agricultural Electrification.'' *Ching-chi Yen-chiu,* no. 3 (March 17), 1963. SCMM 360, p. 2.

Tung Ta-lin. *Agricultural Cooperation in China.* Peking: Foreign Languages Press, 1959.

''Two Diametrically Opposite Lines.'' See: Eighth Ministry of Machine Building, Revolutionary Great Alliance Headquarters and Revolutionary Great Criticism and Repudiation Group of Organizations.

''Use Mao Tse-tung's Thought,'' See: Shansi, Jui-ch'eng County Revolutionary Committee.

Wang Chen. ''Exert Revolution Efforts to Achieve a Rapid Advance in State Farm and Ranch Production.'' Speech to First National Congress of Chinese Agricultural and Water Conservancy Workers' Trade Unions, Jan. 30, 1958. *Kung-jen Jih-pao,* Feb. 1, 1958. SCMP 1723, p. 2.

——. ''Strengthening the Construction of State Farms.'' *Hung-ch'i,* no. 7 (April 1), 1961. SCMM 258, p. 6.

Wang Kuang-wei. ''Actively and Steadily Carry out Technical Transformation of Agriculture.'' *Ching-chi Yen-chiu,* no. 3 (March 17), 1963. SCMM 361, p. 34.

''Wipe Out State Monopoly.'' See: Eighth Ministry of Machine Building, United Committee of the Revolutionary Rebels, ''57'' United Detachment.

Secondary Sources

Barnett, A. Doak. *Cadres, Bureaucracy, and Political Power.* New York: Columbia University Press, 1967.

Berger, Roland. ''The Mechanisation of Chinese Agriculture.'' *Eastern Horizon* 11:3 (1972), 7–26.

Bernstein, Thomas. ''Leadership and Mass Mobilisation Campaigns of 1929–30 and 1955–56: A Comparison.'' *China Quarterly,* no. 31 (July–Sept. 1967), 1–47.

——. ''Problems of Village Leadership after Land Reform.'' *China Quarterly,* no. 36 (Oct.–Dec. 1968), 1–22.

Bowie, Robert, and John Fairbank. *Communist China, 1955–1959.* Cambridge: Harvard University Press, 1965.

Buck, John. *Land Utilization in China.* Nanking: University of Nanking, 1937.

——. *Land Utilization in China, Statistics.* Chicago: University of Chicago Press, 1937.

Chang, Parris. *Power and Policy in China.* University Park: Pennsylvania State University Press, 1975.

Chao, Kang. *Agricultural Production in Communist China, 1949–1965.* Madison: University of Wisconsin Press, 1970.

Chao Kuo-chün. *Agrarian Policies of Mainland China: A Documentary Study (1949–1956).* Cambridge: Harvard East Asian Research Center, 1957.

——. *Agrarian Policy of the Chinese Communist Party, 1929–1950*. Bombay: Asia Publishing House, 1960.

Chen, Jack. *Inside the Cultural Revolution*. New York: Macmillan, 1975.

Chen, Nai-Ruenn. *Chinese Economic Statistics*. Chicago: Aldine, 1967.

Chi, I'chai. "A Study on the Production and Need of Agricultural Machinery on China Mainland." *Fei-ch'ing Yueh-pao* [Monthly Journal on Bandit Situation] 10:3 (May 31, 1967), 63–72. JPRS 42,271, pp. 1–21.

Child, Frank, and Hiromitsu Kaneda. "Links to the Green Revolution: A Study of Small-Scale, Agriculturally Related Industry in the Pakistan Punjab." *Economic Development and Cultural Change* 23:2 (Jan. 1975), 249–275.

Chin, Szu-kai, and Wing-fai Choa. "The Mechanization of Agriculture." In *Contemporary China, 1962–1964,* edited by E. S. Kirby [selective entry], pp. 1–9. Hong Kong: Hong Kong University Press, 1968.

Current Scene editor. "The Conflict between Mao Tse-tung and Liu Shao-ch'i over Agricultural Mechanization in Communist China." *Current Scene* 6:17 (Oct. 1, 1968), 1–20.

Dobb, Maurice. *Soviet Economic Development since 1917*. New York: International Publishers, 1966.

Donnithorne, Audrey. *China's Economic System*. New York: Praeger, 1967.

Gotsch, Carl. "Technical Change and the Distribution of Income in Rural Areas." *American Journal of Agricultural Economics* 54:2 (May 1972), 326–341.

——. "Tractor Mechanisation and Rural Development in Pakistan." *International Labour Review* 107:2 (Feb. 1973), 133–66.

Grey, Jack. "The Economics of Maoism." *Bulletin of the Atomic Scientists* 25 (Feb. 1969), 42–552.

Harding, Harry. "Political Trends in China since the Cultural Revolution." American Academy of Political and Social Sciences, *Annals* 402 (July 1972), 71.

Hinton, William. *Fanshen*. New York: Monthly Review Press, 1966.

——. *Iron Oxen*. New York: Vintage, 1971.

——. "Progress in Long Bow and a Visit to Tsinghua." *China and U.S.* 5:4 (July–Aug. 1976), p. 16.

History of the Communist Party of the Soviet Union (B). New York: International Publishers, 1939.

Huang, Joe C. *Heroes and Villains in Communist China*. New York: Pica Press, 1973.

Huang, Philip C. C. "Analyzing the Twentieth-Century Countryside: Revolutionaries versus Western Scholarship." *Modern China* 1:2 (April 1975), 147–50.

Jasny, Naum. *The Socialized Agriculture of the USSR*. Stanford: Stanford University Press, 1949.

——. *The Soviet Price System*. Stanford: Stanford University Press, 1951.

Joffe, Ellis. "The PLA in Internal Politics." *Problems of Communism* 21:4 (Nov.–Dec. 1975), 1–12.

Jones, P. H. M. "Machines on the Farm." *Far Eastern Economic Review* 45 (Sept. 10, 1964), 480.

Katz, Zev. "Insights from Emigres and Sociological Studies on the Soviet Union." In *Soviet Economic Prospects for the Seventies*. Washington, D.C.: U.S. Congress Joint Economic Committee, June 27, 1973.

Klein, Donald, and Anne Clark. *Biographic Dictionary of Chinese Communism*. Cambridge: Harvard University Press, 1971.

Kuo, Leslie T. C. "Agricultural Mechanization in Communist China." *China Quarterly,* no. 17 (Jan.–March 1964), 134–151.

——. "Industrial Aid to Agriculture in Communist China." *International Development Review* 9:2 (1967), 6–16, 29.

Laird, Roy. *Collective Farming in Russia*. Lawrence: A University of Kansas Publication, Social Science Studies, 1958.

——. Darwin Sharp, and Ruth Sturtevant. *The Rise and Fall of the MTS as an Instrument of Soviet Rule*. Lawrence: Governmental Research Center, University of Kansas, 1960.

Lee, Peter Nan-shong. "Authority in Chinese Industrial Bureaucracy: Recent Developments." Paper for American Society for Public Administration, 1974.

Liberman, E. G. *Economic Methods and the Effectiveness of Production*. New York: Anchor, 1973.

Lin, Chen. "The Agricultural Plight of the Chinese Communists." *Issues and Studies* 5:5 (Feb. 1970), 41–42.

Lippit, Victor. "The Great Leap Forward Reconsidered." *Modern China* 1:1 (Jan. 1975), 92–115.

——. *Land Reform and Economic Development in China*. White Plains: International Arts and Sciences Press, 1974.

"Mass Murder in Communist China." *American Federation of Labor,* Free Trade Union Committee *News* 7:12 (Dec. 1952), 1 ff.

Miller, Robert. *One Hundred Thousand Tractors—The MTS and the Development of Controls in Soviet Agriculture*. Cambridge: Harvard University Press, 1970.

Morawetz, David. "Employment Implications of Industrialization in Developing Countries: A Survey." *Economic Journal* 84 (1974), 491–542.

Nickum, Jim. "A Collective Approach to Water Resource Development: The Chinese Commune System, 1962–72." Ph.D. dissertation, University of California at Berkeley, 1974.

Nicolaus, Martin. *Restoration of Capitalism in the USSR*. Chicago: Liberator Press, 1975.

Oksenberg, Michel. "Policy Formulation in Communist China: The Case of the Mass Irrigation Campaign, 1957–58." Ph.D. dissertation, Columbia University, 1969.

——. "Policy Making under Mao, 1948–68: An Overview." In *China: 'Management of a Revolutionary Society,* edited by John Lindbeck. Seattle: University of Washington Press, 1971.

Pan Hong-sheng and O. T. King. "Preliminary Note on an Economic Study of

Farm Implements.'' *Economic Facts* (University of Nanking, College of Agriculture and Forestry), no. 3 (Nov. 1936), 155.

Pan Hong-sheng and John Raeburn ''Ownership and Costs of Farm Implements and Work Animals in Suhsien, Anhwei.'' *Economic Facts* (University of Nanking, College of Agriculture and Forestry), no. 7 (Oct. 1937), 300.

Perkins, Dwight. ''China's Fourth Five-Year Plan.'' *Current Scene* 12: 9 (Sept. 1974), 2.

——. ''Constraints Influencing China's Agricultural Performance.'' In *China: A Reassessment of the Economy*. Washington, D.C.: U.S. Congress Joint Economic Committee, 1975.

——, ed. *Rural Small-Scale Industry in the People's Republic of China*. Berkeley: University of California Press, 1977.

Porter, D. Gareth. ''The Myth of the Bloodbath: North Vietnam's Land Reform Reconsidered.'' Cornell University International Relations of East Asia Project, Interim Report no. 2, 1972. Summarized in *Indochina Chronicle,* no. 19 (Sept. 15, 1972), 1–5.

Riskin, Carl. ''Small Industry and the Chinese Model of Development.'' *China Quarterly,* no. 46 (April–June 1971), 245–273.

Schram, Stuart, ed. *Chairman Mao Talks to the People*. New York: Pantheon, 1974.

Schran, Peter. *The Development of Chinese Agriculture, 1950–1959*. Urbana: University of Illinois Press, 1969.

Schroeder, Gertrude. ''Consumer Problems and Prospects.'' *Problems of Communism* 22: 2 (March–April 1973), 19.

——. ''Recent Developments in Soviet Planning and Incentives.'' In *Soviet Economic Prospects for the Seventies,* pp. 11–38. Washington, D.C.: U.S. Congress Joint Economic Committee, June 27, 1973.

——. ''Soviet Economic Reform at an Impasse.'' *Problems of Communism* 20:4 (July–Aug. 1971), 36.

——. ''Soviet Technology: System vs. Progress.'' *Problems of Communism* 19:5 (Sept.–Oct. 1970), 19–30.

Schurmann, Franz. *Ideology and Organization in Communist China*. Berkeley: University of California Press, 1968.

Shue, Vivienne. ''Reorganizing Rural Trade: Unified Purchase and Socialist Transformation.'' *Modern China* 2:1 (Jan. 1976), 104–134.

——. ''Taxation, 'Hidden Land,' and the Chinese Peasant.'' *Peasant Studies Newsletter,* Oct. 1974, pp. 1–12.

——. ''Transforming China's Peasant Villages: Rural Political and Economic Organization, 1949–56.'' Ph.D. dissertation, Harvard University, 1975.

Sigurdson, Jon. ''Rural Industrialization in China.'' In *China: A Reassessment of the Economy,* pp. 411–435. Washington, D.C.: U.S. Congress Joint Economic Committee, 1975.

——. ''Rural Industry and the Internal Transfer of Technology.'' In *Authority, Participation and Cultural Change in China,* edited by Stuart Schram, pp. 199–232. Cambridge: Cambridge University Press, 1973.

——. "Rural Industry—A Traveller's View." *China Quarterly,* no. 50 (April–June 1972), 315–337.

Sladkovskii, M. I. *History of Economic Relations between Russia and China.* Translation by Israel Program for Scientific Translations, Jerusalem, 1966. Reported by NCNA, Jan. 9, 1956. SCMP 1203. Originally published Moscow, 1957.

Southworth, Herman, ed. *Farm Mechanization in East Asia.* New York: Agricultural Development Council, 1972.

Soviet Economic Reform: Progress and Problems. Moscow: Progress Publishers, 1972.

Stalin, J. *Economic Problems of Socialism.* Moscow: Foreign Press, Press 1952.

Stavis, Benedict. *Making Green Revolution.* Ithaca: Cornell Rural Development Committee, 1974.

——. "A Preliminary Model for China's Grain Production, 1974." *China Quarterly,* no. 65 (March 1976), 82–96.

Strong, Anna Louise. *The Rise of the People's Communes in China.* New York: Marzani and Munsell, 1960.

Tannebaum, Gerald. *The Real Spirit of Tachai,* New York: MSS Publications, 1974.

Thomas, John Woodward. "The Choice of Technology for Irrigation Tubewells in East Pakistan: An Analysis of a Development Policy Decision." In *The Choice of Technology in Developing Countries (Some Cautionary Tales),* edited by Peter Timmer et al., pp. 41–50. Cambridge: Harvard Center for International Affairs, 1975.

United Nations Relief and Rehabilitation Administration. *Agricultural Rehabilitation in China.* Operational Analysis Papers, No. 52. Washington, D.C.: UNRRA, April 1948.

Vail, David. "The Case for Rural Industry: Economic Factors Favoring Small Scale, Decentralized, Labor-Intensive Manufacturing." Unpublished paper, July 1975.

——. *Technology for Ujamaa Village Development in Tanzania.* Syracuse University, Foreign and Comparative Studies, Eastern Africa Series 18, 1975.

Volin, Lazar. *A Century of Russian Agriculture.* Cambridge: Harvard University Press, 1970.

Walker, Kenneth. "Collectivisation in Retrospect: The Socialist High Tide of Autumn 1955–Spring 1956." *China Quarterly,* no. 26 (April–June 1966), 1–43.

——. "Organization of Agricultural Production." In *Economic Trends in Communist China,* edited by Alexander Eckstein, Walter Galenson, and Ta-chung Liu. Chicago: Aldine, 1968.

Walker, Richard. *The Human Cost of Communism in China.* Washington, D.C.: Government Printing Office, 1971.

Williams, Bobby. "The Chinese Petroleum Industry: Growth and Prospects." In *China: A Reassessment of the Economy.* Washington, D.C.: U.S. Congress Joint Economic Committee, 1975.

Wong, John. *Land Reform in the People's Republic of China*. New York: Praeger, 1973.
——. "Peasant Economic Behavior: The Case of Traditional Agricultural Cooperation in China." *Developing Economies* 9 (1971), 332–349.

Visitors to China whose Unpublished Diaries Were Consulted

Ian Davies (1968)
Arthur Galston (1971)
Frank Kehl (1971)
Bruce McFarland (1968)
Ben Stavis (1972)

Chinese Periodicals Cited

Cheng-chih Hsüeh-hsi [Political Study]
Chi-hsieh Kung-yeh [Machine Building Industry]
Chi-hua Ching-chi [Planned Economy]
Chiang-su Nung-hsüeh-pao [Kiangsu Agricultural Journal]
Chieh-fang Jih-pao [Liberation Daily]
China Reconstructs
Ching-chi Yen-chiu [Economic Research]
Ch'iu-shih [Seeking the Truth]
Chung-kuo Ch'ing-nien Pao [China Youth Newspaper]
Chung-kuo Hsin-wen [Chinese News]
Chung-kuo Nung-pao [Chinese Agricultural News]
Chung-kuo Nung-yeh [Chinese Agriculture]
Chung-kuo Nung-yeh Chi-hsieh [Chinese Agricultural Machinery]
Chung-kuo Nung-yeh K'o-hsüeh [Chinese Agricultural Science]
Current Digest of the Soviet Press
Honan Jih-pao [Honan Daily]
Hsin-hua Yüeh-pao [New China Monthly]
Hsüeh-hsi [Study]
Hsüeh K'o-hsüeh [Study Science]
Hung-ch'i [Red Flag]
Jen-min Jih-pao [People's Daily]
K'o-hsüeh T'ung-pao [Science Report]
Kuang-ming Jih-pao [Bright Daily]
Kuang-tung Nung-yeh K'o-hsüeh [Kwangtung Agricultural Science]
Kung-jen Jih-pao [Worker's Daily]
Liang Shih [Grain]
NCNA [New China News Agency] news releases
Nan-fang Jih-pao [Southern Daily]
Nung-yeh Chi-hsieh [Agricultural Machinery]
Nung-yeh Chi-hsieh Chi-shu [Agricultural Machinery Technique]
Nung-yeh Chi-shu [Agricultural Technique]
Peking Review
Shih-shih Shou-tse [Handbook of Facts]

Ta Kung Pao [Impartial Newspaper]
T'ung-chi Kung-tso [Statistical Work]
T'ung-chi Kung-tso T'ung-hsùn [Statistical Work Bulletin]
Yang-ch'eng Wan-pao [Evening News], Canton

Translation Services Used

CB	Current Background (U.S. Consulate, Hong Kong)
ECMM	Extracts from China Mainland Magazine (U.S. Consulate Hong Kong)
FBIS	Foreign Broadcast Information Service (U.S. Government)
JPRS	Joint Publications Research Service (U.S. Department of Commerce)
SCMM	Survey of China Mainland Magazine (U.S. Consulate, Hong Kong)
SCMP	Survey of China Mainland Press (U.S. Consulate, Hong Kong)
URS	Union Research Service (Union Research Institute, Hong Kong)

Index

Library of Congress Cataloging in Publication Data
(For library cataloging purposes only)
Stavis, Ben.
The politics of agricultural mechanization in China.
Bibliography: p.
Includes index.
1. Agriculture and state—China. 2. Farm mechanization—China. I. Title.
HD2098 1978.S7 1978 338.1'851 77-90916
ISBN 0-8014-1087-8